Defoe Abbott Marx Hardy Machiavelli Chesterton Emerson Joyce Austen
Melville Montaigne Cooper Hugo
Haggard Eliot Grimm
Stoker Christie Molière
Wilde Carroll Maupassant Byron Schiller
Garnett Fitzgerald Engels Smith Kafka
Goethe Einstein Hawthorne Hall
Cotton Dostoyevsky
Baum Kipling Doyle Willis
Leslie Dumas Henry Nietzsche
Flaubert Turgenev Balzac
Stockton Vatsyayana Crane
Burroughs Verne
Curtis Tocqueville Gogol Vinci
Homer Widger Whitman Busch
Darwin Tolstoy
Potter Freud Thoreau Twain Scott
Zola Plato Harte
Kant Jowett Lawrence Hesse
Stevenson Dickens Burton
Andersen Cervantes
London Descartes Voltaire
Poe Aristotle Wells
Hale James Hastings Cooke
Bunner Shakespeare Irving
Richter Chambers
Doré da Benedict
Dante Shaw Pushkin Alcott
Swift Chekhov Wodehouse
Newton

tredition

tredition was established in 2006 by Sandra Latusseck and Soenke Schulz. Based in Hamburg, Germany, tredition offers publishing solutions to authors and publishing houses, combined with worldwide distribution of printed and digital book content. tredition is uniquely positioned to enable authors and publishing houses to create books on their own terms and without conventional manufacturing risks.

For more information please visit: www.tredition.com

TREDITION CLASSICS

This book is part of the TREDITION CLASSICS series. The creators of this series are united by passion for literature and driven by the intention of making all public domain books available in printed format again - worldwide. Most TREDITION CLASSICS titles have been out of print and off the bookstore shelves for decades. At tredition we believe that a great book never goes out of style and that its value is eternal. Several mostly non-profit literature projects provide content to tredition. To support their good work, tredition donates a portion of the proceeds from each sold copy. As a reader of a TREDITION CLASSICS book, you support our mission to save many of the amazing works of world literature from oblivion. See all available books at www.tredition.com.

Project Gutenberg

The content for this book has been graciously provided by Project Gutenberg. Project Gutenberg is a non-profit organization founded by Michael Hart in 1971 at the University of Illinois. The mission of Project Gutenberg is simple: To encourage the creation and distribution of eBooks. Project Gutenberg is the first and largest collection of public domain eBooks.

Freie wissenschaft und freie lehr. English

Ernst Heinrich Philipp August Haeckel

Imprint

This book is part of TREDITION CLASSICS

Author: Ernst Heinrich Philipp August Haeckel
Cover design: Buchgut, Berlin – Germany

Publisher: tredition GmbH, Hamburg - Germany
ISBN: 978-3-8472-1542-4

www.tredition.com
www.tredition.de

Copyright:
The content of this book is sourced from the public domain.

The intention of the TREDITION CLASSICS series is to make world literature in the public domain available in printed format. Literary enthusiasts and organizations, such as Project Gutenberg, worldwide have scanned and digitally edited the original texts. tredition has subsequently formatted and redesigned the content into a modern reading layout. Therefore, we cannot guarantee the exact reproduction of the original format of a particular historic edition. Please also note that no modifications have been made to the spelling, therefore it may differ from the orthography used today.

FREEDOM IN SCIENCE AND TEACHING.

FREEDOM

IN

SCIENCE AND TEACHING.

FROM THE GERMAN OF

ERNST HAECKEL.
WITH A PREFATORY NOTE

By T. H. HUXLEY, F.R.S.

Der Teleolog
"Welche Verehrung verdient der Weltenschöpfer der gnädig.
Als er den Korkbaum schuf, gleich auch die Stöpfel erfand."
Xenien.

NEW YORK:
D. APPLETON AND COMPANY,
549 AND 551 BROADWAY.
1879.

PREFATORY NOTE.

In complying with the wish of the publishers of Professor Haeckel's [v] reply to Professor Virchow, that I should furnish a prefatory note expressing my own opinion in respect of the subject-matter of the controversy, Gay's homely lines, prophetic of the fate of those "who in quarrels interpose," emerge from some brain-cupboard in which they have been hidden since my childish days. In fact, the hard-hitting with which both the attack and the defence abound, makes me think with a shudder upon the probable sufferings of the unhappy man whose intervention should lead two such gladiators to turn their weapons from one another upon him. In my youth, I once attempted to stop a street fight, and I have never forgotten the brief but impressive lesson on the value of the policy of non-intervention which I then received.

But there is, happily, no need for me to place myself in a position which, besides being fraught with danger, would savour of presumption: [vi] Careful study of both the attack and the reply leaves me without the inclination to become either a partisan or a peacemaker: not a partisan, for there is a great deal with which I fully agree said on both sides; not a peacemaker, because I think it is highly desirable that the important questions which underlie the discussion, apart from the more personal phases of the dispute, should be thoroughly discussed. And if it were possible to have controversy without bitterness in human affairs, I should be disposed, for the general good, to use to both of the eminent antagonists the famous phrase of a late President of the French Chamber— "*Tape dessus.*"

No profound acquaintance with the history of science is needed to produce the conviction, that the advancement of natural knowledge has been effected by the successive or concurrent efforts of men, whose minds are characterised by tendencies so opposite that they are forced into conflict with one another. The one intellect is imaginative and synthetic; its chief aim is to arrive at a broad and coherent conception of the relations of phenomena; the other is positive, critical, analytic, and sets the highest value upon the exact determination and statement of the phenomena themselves.

If the man of the critical school takes the pithy aphorism "Melius [vii] autem est naturam secare quam abstrahere" [1] for his motto, the champion of free speculation may retort with another from the same hand, "Citius enim emergit veritas e falsitate quam e confusione;" [2] and each may adduce abundant historical proof that his method has contributed as much to the progress of knowledge as that of his rival. Every science has been largely indebted to bold, nay, even to wild hypotheses, for the power of ordering and grasping the endless details of natural fact which they confer; for the moral stimulus which arises out of the desire to confirm or to confute them; and last, but not least, for the suggestion of paths of fruitful inquiry, which, without them, would never have been followed. From the days of Columbus and Kepler to those of Oken, Lamarck, and Boucher de Perthes, Saul, who, seeking his father's asses, found a kingdom, is the prototype of many a renowned discoverer who has lighted upon verities while following illusions, which, had they deluded lesser men, might possibly have been considered more or less asinine.

On the other hand, there is no branch of science which does not owe at least an equal obligation to those cool heads, which are not to be seduced into the acceptance of symmetrical formulæ and bold generalisations for solid truths because of their brilliancy [viii] and grandeur; to the men who cannot overlook those small exceptions and insignificant residual phenomena which, when tracked to their causes, are so often the death of brilliant hypotheses; to the men, finally, who, by demonstrating the limits to human knowledge which are set by the very conditions of thought, have warned mankind against fruitless efforts to overstep those limits.

Neither of the eminent men of science, whose opinions are at present under consideration, can be said to be a one-sided representative either of the synthetic or of the analytic school. Haeckel, no less than Virchow, is distinguished by the number, variety, and laborious accuracy of his contributions to positive knowledge; while Virchow, no less than Haeckel, has dealt in wide generalisations, and, until the obscurantists thought they could turn his recent utterances to account, no one was better abused by them as a typical freethinker and materialist. But, as happened to the two women grinding at the same mill, one has been taken and the other left. Since the publication of his famous oration, Virchow has been received into the bosom of orthodoxy and respectability, while Haeckel remains an outcast!

To those who pay attention to the actual facts of the case, this is a very surprising event; and I confess that nothing has ever perplexed me more than the reception [ix] which Professor Virchow's oration has met with, in his own and in this country; for it owes that reception, not to the undoubted literary and scientific merits which it possesses, but to an imputed righteousness for which, so far as I can discern, it offers no foundation. It is supposed to be a recantation; I can find no word in it which, if strictly construed, is inconsistent with the most extreme of those opinions which are commonly attributed to its author. It is supposed to be a deadly blow to the doctrine of evolution; but, though I certainly hold by that doctrine with some tenacity, I am able, *ex animo*, to subscribe to every important general proposition which its author lays down.

In commencing his address, Virchow adverts to the complete freedom of investigation and publication in regard to scientific questions which obtains in Germany; he points out the obligation which lies upon men of science, even if for no better reason than the maintenance of this state of things, to exhibit a due sense of the responsibility which attaches to their speaking and writing, and he dwells on the necessity of drawing a clear line of demarcation between those propositions which they have a fair right to regard as established truths, and those which they know to be only more or less well-founded speculations. Is any one prepared to deny that this is the first great commandment of the [x] ethics of teaching? Would any responsible scientific teacher like to admit that he had not done his best to separate facts from hypotheses in the minds of his hearers; and that he had not made it his chief business to enable those whom he instructs to judge the latter by their knowledge of the former?

More particularly does this obligation weigh upon those who address the general public. It is indubitable, as Professor Virchow observes, that "he who speaks to, or writes for, the public is doubly bound to test the objective truth of that which he says." There is a sect of scientific pharisees who thank God that they are not as those publicans who address the public. If this sect includes anybody who has attempted the business without failing in it, I suspect that he must have given up keeping a conscience. For assuredly if a man of science, addressing the public, bethinks him, as he ought to do, that the obligation to be accurate—to say no more than he has warranty for, without clearly marking off so much as is hypothetical—is far heavier than if he were dealing with experts, he will find his task a very admirable mental exercise. For my own part, I am inclined to doubt whether there is any method of self-discipline better calculated to clear up one's own ideas about a difficult subject, than that which arises out of the effort to put them forth, with fulness and precision, in language [xi] which all the world can understand. Sheridan is said to have replied to some one who remarked on the easy flow of his style, "Easy reading, sir, is—hard writing;" and any one who is above the level of a scientific charlatan will know that easy speaking is "— —hard thinking."

Again, when Professor Virchow enlarges on the extreme incompleteness of every man's knowledge beyond those provinces which he has made his own (and he might well have added within these also), and when he dilates on the inexpediency, in the interests of science, of putting forth as ascertained truths propositions which the progress of knowledge soon upsets—who will be disposed to gainsay him? Nor have I, for one, anything but cordial assent to give to his declaration, that the modern development of science is essentially due to the constant encroachment of experiment and observation on the domain of hypothetical dogma; and that the most difficult, as well as the most important, object of every honest worker is *"sich ent-subjectiviren"*—to get rid of his preconceived notions, and to keep his hypotheses well in hand, as the good servants and bad masters that they are.

I do not think I have omitted any one of Professor Virchow's main theses in this brief enumeration. I do not find that they are disputed by Haeckel, and [xii] I should be profoundly astonished if they were. What, then, is all the coil about, if we leave aside various irritating sarcasms, which need not concern peaceable Englishmen?

Certainly about nothing that touches the present main issues of scientific thought. The "plastidule-soul" and the potentialities of carbon may be sound scientific conceptions, or they may be the reverse, but they are no necessary part of the doctrine of evolution, and I leave their defence to Professor Haeckel.

On the question of equivocal generation, I have been compelled, more conspicuously and frequently than I could wish, during the last ten years, to enunciate exactly the same views as those put forward by Professor Virchow; so that, to my mind, at any rate, the denial that any such process has as yet been proved to take place in the existing state of nature, as little affects the general doctrine. [3]

With respect to another side issue, raised by Professor Virchow, he appears to me to be entirely in the wrong. He is careful to say that he has no [xiii] unwillingness to accept the descent of man from some lower form of vertebrate life; but, reminding us of the special attention which, of late years, he has given to anthropology, he affirms that such evidence as exists is not only insufficient to support that hypothesis, but is contrary to it. "Every positive progress which we have made in the region of prehistoric anthropology has removed us further from the demonstration of this relation."

Well, I also have studied anthropological questions in my time; and I feel bound to remark, that this assertion of Professor Virchow's appears to me to be a typical example of the kind of incautious over-statement which he so justly reprehends.

For, unless I greatly err, all the real knowledge which we possess of the fossil remains of man goes no farther back than the Quaternary epoch; and the most that can be asserted on Professor Virchow's side respecting these remains is, that none of them present us with more marked pithecoid characters than such as are to be found among the existing races of mankind. [4] But, if this be so, then the only just conclusion to be drawn from the evidence as it stands is, that the men of the Quaternary epoch may have proceeded [xiv] from a lower type of humanity, though their remains hitherto discovered show no definite approach towards that type. The evidence is not inconsistent with the doctrine of evolution, though it does not help it. If Professor Virchow had paid as much attention to comparative anatomy and palæontology as he has to anthropology, he would, I doubt not, be aware that the equine quadrupeds of the Quaternary period do not differ from existing *Equidæ* in any more important respect than these last differ from one another; and he would know that it is, nevertheless, a well-established fact that, in the course of the Tertiary period, the equine quadrupeds have undergone a series of changes exactly such as the doctrine of evolution requires. Hence sound analogical reasoning justifies the expectation that, when we obtain the remains of Pliocene, Miocene, and Eocene *Anthropidæ*, they will present us with the like series of gradations, notwithstanding the fact, if it be a fact, that the Quaternary men, like the Quaternary horses, differ in no essential respect from those which now live.

I believe that the state of our knowledge on this question is still justly summed up in words written some seventeen years ago: —

"In conclusion, I may say, that the fossil remains of man hitherto discovered do not seem to me to take us appreciably nearer to that lower pithecoid form by [xv] the modification of which he has probably become what he is. And considering what is now known of the most ancient races of men; seeing that they fashioned flint axes, and flint knives, and bone skewers of much the same pattern as those fabricated by the lowest savages at the present day, and that we have every reason to believe the habits and modes of living of such people to have remained the same from the time of the mammoth and the tichorhine rhinoceros till now, I do not know that the result is other than might be expected." [5]

I have seen no reason to change the opinion here expressed, and so far from the fact being in the slightest degree opposed to a belief in the evolution of man, all that has been learned of late years respecting the relation of the Recent and Quaternary to the Tertiary mammalia appears to me to be in striking harmony with what we know respecting Quaternary man, supposing man to have followed the general law of evolution.

The only other collateral question of importance raised by Professor Virchow is, whether the doctrine of evolution should be generally taught in schools or not. Now I cannot find that Professor Virchow anywhere distinctly repudiates the doctrine; all that he distinctly says is that it is not proven, and that things which [xvi] are not proven should not be authoritatively instilled into the minds of young people.

If Professor Virchow will agree to make this excellent rule absolute, and applicable to all subjects that are taught in schools, I should be disposed heartily to concur with him.

But what will his orthodox allies say to this? If "not provenness" is susceptible of the comparative degree, by what factor must we multiply the imperfection of the evidence for evolution in order to express that of the evidence for special creation; or to what fraction must the value of the evidence in favour of the uninterrupted succession of life be reduced in order to express that in support of the deluge? Nay, surely even Professor Virchow's "dearest foes," the "plastidule soul" and "Carbon & Co.," have more to say for themselves, than the linguistic accomplishments of Balaam's ass and the obedience of the sun and moon to the commander of a horde of bloodthirsty Hebrews! But the high principles of which Professor Virchow is so admirable an exponent do not admit of the application of two weights and two measures in education; and it is surely to be regretted that a man of science of great eminence should advocate the stern bridling of that teaching which, at any rate, never outrages common sense, nor refuses to submit to criticism, while he has no whisper of remonstrance [xvii] to offer to the authoritative propagation of the preposterous fables by which the minds of children are dazed and their sense of truth and falsehood perverted. Professor Virchow solemnly warns us against the danger of attempting to displace the Church by the religion of evolution. What this last confession of faith may be I do not know, but it must be bad indeed if it inculcates more falsities than are at present foisted upon the young in the name of the Church.

I make these remarks simply in the interests of fair play. Far be it from me to suggest that it is desirable that the inculcation of the doctrine of evolution should be made a prominent feature of general education. I agree with Professor Virchow so far, but for very different reasons. It is not that I think the evidence of that doctrine insufficient, but that I doubt whether it is the business of a teacher to plunge the young mind into difficult problems concerning the origin of the existing condition of things. I am disposed to think that the brief period of school-life would be better spent in obtaining an acquaintance with nature, as it is; in fact, in laying a firm foundation for the further knowledge Which is needed for the critical examination of the dogmas, whether scientific or anti-scientific, which are presented to the adult mind. At present, education proceeds in the reverse way; the teacher makes the most confident assertions on precisely those subjects [xviii] of which he knows least; while the habit of weighing evidence is discouraged, and the means of forming a sound judgment are carefully withheld from the pupil.

Professor Virchow is known to me only as he is known to the world in general—by his high and well-earned scientific reputation. With Professor Haeckel, on the other hand, I have the good fortune to be on terms of personal friendship. But in making the preceding observations, I should be sorry to have it supposed that I am holding a brief for my friend, or that I am disposed to adopt all the opinions which he has expressed in his reply. Nevertheless, I do desire to express my hearty sympathy with his vigorous defence of the freedom of learning and teaching; and I think I shall have all fair-minded men with me when I also give vent to my reprobation of the introduction of the sinister arts of unscrupulous political warfare into scientific controversy, manifested in the attempt to connect the doctrines he advocates with those of a political party which is, at present, the object of hatred and persecution in his native land. The one blot, so far as I know, on the fair fame of Edmund Burke is his attempt to involve Price and Priestley in the furious hatred of the English masses against the authors and favourers of the revolution of [xix] 1789. Burke, however, was too great a man to be absurd, even in his errors; and it is not upon record that he asked uninformed persons to consider what might be the effect of such an innovation as the discovery of oxygen on the minds of members of the Jacobin Club.

Professor Virchow is a politician—maybe a German Burke, for anything that I know to the contrary; at any rate, he knows the political value of words; and, as a man of science, he is devoid of the excuses that might be made for Burke. Nevertheless, he gravely charges his hearers to "imagine what shape the theory of descent takes in the head of a Socialist."

I have tried to comply with this request, but I have utterly failed to call up the dread image; I suppose because I do not sufficiently sympathise with Socialists. All the greater is my regret that Professor Virchow did not himself unfold the links of the hidden bonds which unite evolution with revolution, and bind together the community of descent with the community of goods.

Professor Virchow is, I doubt not, an accomplished English scholar. Let me commend the "Rejected Addresses" to his attention. For since the brothers Smith sang—

"Who makes the quartern loaf and Luddites rise,"—
Who fills the butchers' shops with large blue flies,

there has been nothing in literature at all comparable [xx] to the attempt to frighten sober people by the suggestion that evolutionary speculations generate revolutionary schemes in Socialist brains. But then the authors of the "Rejected Addresses" were joking, while Professor Virchow is in grim earnest; and that makes a great difference in the moral aspect of the two achievements.

[1] Novum Organon, li.

[2] Partis instaurationis secundæ delineatio.

[3] I may remark parenthetically that Professor Virchow's statement of the attitude of Harvey towards equivocal generation is strangely misleading. For Harvey, as every student of his works knows, believed in equivocal generation; and, in the sense in which he uses the word ovum, "nempe substantiam quandam corpoream vitam habentem potentia," the truth of the axiom "omne vivum ex ovo," popularly ascribed to him, has in no wise been affected by the discoveries of later days in the manner asserted by Professor Virchow.

[4] I do not admit that so much can be said; for the like of the Neanderthal skull has yet to be produced from among the crania of existing men.

[5] Man's Place in Nature, p. 159.

[xxi]

PREFACE.

When the address delivered by Rudolph Virchow on the 22d of September last year, at the fiftieth meeting of German Naturalists and Physicians at Munich, on "Freedom of Science in the Modern State," appeared in print in the following October, I was called upon, on many sides, to prepare a reply. And such a reply on my part seemed, in fact, justified by the severe strictures which Virchow in his discourse had directed against one delivered by me only four days previously, before the same meeting, on "The Modern Doctrine of Evolution in its Relation to General Science." The general views which Virchow then unfolded proved such a fundamental opposition in our principles, and touched our dearest moral convictions so nearly, that any reconciliation of such antagonistic views was no longer to be thought of. Nevertheless I forbore publishing the ready reply for two reasons: one relating to the matter itself, the other a personal one.

With regard to the matter itself, I believed I might confidently leave it to futurity to decide in the contention [xxii] that has declared itself between us. For on one hand the doctrine of evolution which Virchow attacks has already so far become a sure basis of biological science and part of the most precious mental-stock of cultivated humanity, that neither the anathemas of the Church nor the contradiction of the greatest scientific authority—and such an one is Virchow—can prevail against it; and on the other hand most of the arguments which he specially adduces against the theory of descent have been so often discussed and so thoroughly refuted that any renewed discussion seems in fact superfluous.

Personally, it was in the highest degree repugnant to me to come forward as the opponent of a man whom I learned, a quarter of a century ago, to acknowledge and to honour as the reformer of medical science; a man whose most ardent disciple and most enthusiastic follower I at that time was, with whom I subsequently stood in the closest relation as his assistant, and with whom I long after continued in the most friendly intercourse. The more keenly I lamented Virchow's position, for some years past, as the antagonist of our modern doctrine of evolution, and the more I felt myself challenged to a reply by his repeated attacks upon it, the less inclination I felt, nevertheless, to come forward publicly as the opponent of this distinguished and highly-honoured man. [xxiii]

And if I find myself, after all, forced to reply, it is in the persuasion that a longer silence will add to the erroneous conclusions which my hitherto resigned attitude has already given rise to; at the same time I believe that, precisely by reason of the peculiar interest with which I have throughout followed Virchow's scientific achievements, I am specially qualified to answer the question, a hundred times repeated by letter or by word of mouth—"How is it possible that a man who so long stood at the head of a party of progress in science as in politics, who in political life indeed, has outwardly maintained this position, has in science become an instrument of the most perilous reaction?"

A verbal answer, which I incidentally gave in March of last year at the Concordia Banquet at Vienna, was reported in the daily papers in such a different sense, and was in part so misunderstood or so intentionally misrepresented, that I am forced at last, on that account, to publish a clear and unambiguous reply. The "Augsburger Allgemeine Zeitung," which eagerly seizes every opportunity of expressing its unconquerable aversion to the evolution theory, accused me, in one of its hostile articles, of a virulent and undignified attack on Virchow. In contradiction of this misrepresentation in the Augsburg paper—which was copied by other journals—I must expressly assert that not Virchow but I myself am the person attacked, and [xxiv] that, therefore, the matter in question is not an unjustifiable attack by me on a formerly revered friend, but a defence to which I am compelled by repeated and sharp attacks on his part.

Another reason which urges me at last to break silence consists in the continual and ample advantage that all the clerical and reactionary organs have been taking of Virchow's address, during the last three-quarters of a year, in favour of mental retrogression. The shouts of triumph with which they at once hailed Virchow's "grand moral action," that is to say, his perversion from a Free-thinker to the side of mental darkness, was the first signal for that persistent utilisation of his authority of which the pernicious consequences can by no means be escaped. Friedrich von Hellwald, in his discussion on the speeches made at Munich, has already strikingly pointed out [6] the grave danger that exists when just such an one as Virchow, standing under the banner of political liberalism and wrapped in the mantle of severe science, decisively combats against the freedom of science and of its doctrines. This serious danger has never shown so threatening an aspect as at the present moment, when our political and religious life appears to be encountering such a reaction as has not occurred for a long time. The two insane attempts which, within a few [xxv] weeks, have been made by Social-democracy against the revered and reverend person of the German Emperor have raised a storm of righteous indignation of such violence that calm judgment is entirely overthrown, and that many even of the most liberal of liberal politicians not only impetuously urge us to the severest measures against the Utopian doctrines of social democracy but, far over-shooting the mark, demand that free-doctrine and free-thought, that freedom of the press and even freedom of conscience shall be thrown into the narrowest fetters. Can this reaction, lurking in the background, find any more welcome support than is afforded by the mere demand of such a man as Virchow for restriction of liberty in teaching? And if he makes our present doctrines of evolution in general and the theory of descent in particular responsible for the mad doctrines of social-democracy, it is but a natural and just consequence when the famous New-Prussian "Kreuz-Zeitung" throws all the blame of these treasonable attempts of the democrats Hödel and Nobiling—as in fact it quite lately did—directly on the theory of descent, and especially on the hated doctrine of the "descent of man from apes." And the danger which threatens us shows a still graver aspect when we consider how great an influence Virchow has at the present day as an advanced liberal, and how he is regarded in the [xxvi Prussian diet as the highest practical authority, and at the same time as the most liberal critic when educational questions are under consideration. Now it is well known that one of the most important problems lying before the Prussian parliament is the consideration of a new education-law, which will probably exercise its restricting influence for a long time to come, not in Prussia only, but throughout Germany; what can we expect of such an education-law if in the course of the deliberations, among the small number of those specialists who are generally listened to, Virchow raises his voice as a leading authority, and brings forward the principles that he proclaimed in his speech at Munich as the surest guarantees for the freedom of science in the modern polity? Article XX. of the Prussian Charter, and § 152 of the Code of the German Empire, say, "Science and its doctrines are free." And Virchow's first step, according to the principles he now declares, must be a motion to abrogate this paragraph.

In the face of this imminent danger, I dare no longer hesitate about my answer. *Amicus Socrates, amicus Plato, magis amica Veritas.* An unreserved and public opposition can be no longer postponed. As a matter of fact, at the Munich meeting, neither did Virchow hear my speech nor I his. I read my paper, as it is printed, on the 18th September 1877, and left [xxvii] on the 19th. Virchow came to Munich only on the 20th, and delivered his speech on the 22d.

Bearing in mind the gratitude which I owe to Virchow as my former master and friend at Würzburg — a gratitude which I have at all times striven to prove by the further development of his mechanical theory — I shall confine myself, as far as possible, to an objective and special confutation of his assertions. Certainly the temptation on this occasion was a strong one to pay the debt in like kind. In my Munich lecture, among the few names to which I alluded, I particularly mentioned that of Virchow as the distinguished founder of cellular-pathology (p. 12). [7] Virchow's return for this was to heap scorn and ridicule on the doctrine of evolution in his usual manner. The critic in the "National-Zeitung," Herr Isidor Kastan, says of this with particular satisfaction, "The ridicule with which Herr Virchow treated this side of Haeckel's visions was indeed caustic enough, but this is ever Virchow's way; only in this case, if in any, he was fully justified."

I could less easily ignore Virchow's denunciation of me than his satire — a denunciation which gibbeted me as a confederate in the social-democratic cause, and which made the theory of descent answerable for the horrors of the Paris Commune. The opinion is now widely spread that by this intentional connection of [xxviii] the theory of descent with Social Democracy he has hit the hardest blow at that theory, and that he aimed at nothing less than the removal of all "Darwinists" from their academic chairs and professorships. This is the inevitable consequence of his demands; for if Virchow insists with the utmost determination that the theory of descent must not be taught (because he does not regard it as true), what is to become of the supporters of that theory who, like myself, regard it as incontrovertibly true, and teach it as a perfectly sound theory? And at least nine-tenths of all the teachers of zoology and botany in Europe are among its supporters from immutable conviction of its truth, as well as all morphologists without exception. Virchow cannot expect that these teachers should collectively renounce that which they believe to be immutable truth, and in its place set up the dogma of the Church as the basis of their teaching, in accordance with his wish! Nothing remains for them but to vacate their professors' chairs, and—according to Virchow and the "Germania"—the "Modern Polity" would be in duty bound to deprive them of their liberty of teaching if they did not voluntarily renounce it.

If this be indeed Virchow's purpose, as it is generally supposed to be, with regard to me, at least, he may spare himself the trouble. Amongst us in Jena quite [xxix] other ideas prevail as to the "Freedom of science in the modern Polity" than those which obtain in the capital, Berlin. And among us the Berlin students' rhyme has no meaning,

"Who knows the truth and freely speaks,
On him the law its vengeance wreaks." [8]

The Jena students, on the contrary, sing the rhyme in its original form—

"Who knows the truth and speaks it not,
A feeble wretch is he, God wot." [9]

The Rector Magnificentissimus of the University of Jena, the Grand Duke of Saxony, who has proved himself the protector of the arts and sciences, has besides far more liberal views as to the liberty of scientific investigation and teaching than the illustrious head of the party of progress at Berlin. The enlightened and liberal Prince at Weimar, under whose particular protection we in Jena find ourselves, has never conceived it necessary to limit in any way the unbounded freedom of my teaching and my writing; not even when in 1866 my "General Morphology," and 1868 my "History of Creation" first appeared, and when many people attempted to make the youthful [xxx] extravagances which were to be found in those works the ground of a serious accusation. And what farther mischief have these extravagances done, though I now sincerely lament them?

Faithful to the glorious traditions of a past extending over three centuries, the little Thuringian university of Jena will find a way to preserve her perfect and unlimited freedom. She will ever bear in mind that she is the first Protestant university of Germany, protesting against every strait-waistcoat which hierarchical obstinacy would force upon human reason, against every dogma by which the arrogance of the learned may try to suppress all freedom of teaching. She will freely seek and freely teach in accordance with her highest convictions, untroubled by the fact that in the "great" university of Berlin nothing may be taught, as Virchow insists, but what is objectively ascertained, absolutely sure; that is to say, nothing that rises above individual, indubitable, and intelligible facts; not an idea, not a conception, not a theory, in fact not any real science; mathematics, at most, excepted. It is our conviction that Jena will continue to be an independent city of refuge for free science and free teaching as long as it remains under the faithful nurture and liberal protection of the princely house of Sax Weimar, that enlightened race which is linked with the history of German intellect through the [xxxi] matchless traditions of its glorious past. What the Wartburg was to Martin Luther, what Weimar has been to the foremost heroes of German literature, what Jena herself has been during three hundred years to a vast number of illustrious investigators, that will the tried and tested Jena of to-day undoubtedly continue to be to the modern doctrine of evolution, as to every other doctrine which asks free development; a strong-hold of free thought, free investigation, and free doctrine.

ERNST HAECKEL.
Jena, *June 24th*, 1878.

[6] Kosmos, Vol. II. p. 172.

[7] Of the German.

[8] "Wer die Wahrheit kennet und saget sie frei,
Der kommt in Berlin auf die Stadt-Vogtei."

[9] "Wer die Wahrheit kennet und saget sie nicht
Der ist für wahr ein erbärmlicher Wicht."

CONTENTS.

PAGE

	PREFATORY NOTE	v
	PREFACE	xxi
CHAP.		
I.	DEVELOPMENT AND CREATION	1
II.	CERTAIN PROOFS OF THE DOCTRINE OF DESCENT	10
III.	THE SKULL THEORY AND THE APE THEORY	29
IV.	THE CELL-SOUL AND CELLULAR PSYCHOLOGY	46
V.	THE GENETIC AND DOGMATIC METHODS OF TEACHING	61
VI.	THE DOCTRINE OF DESCENT AND SOCIAL DEMOCRACY	88
VII.	IGNORABIMUS ET RESTRINGAMUR	99

FREEDOM IN SCIENCE AND TEACHING.

[1]

CHAPTER I.

DEVELOPMENT AND CREATION.

Nothing is more helpful for the understanding of scientific controversies, or for the clearing of confused conceptions, than a contrasted statement, as defined and clear as possible, of the simplest leading propositions of the contending doctrines. Hence it is highly favourable to the victory of our modern doctrine of evolution that its chief problem, the question as to the origin of species, is being more and more pressed by these opposite alternatives: Either all organisms are naturally evolved, and must in that case be all descended from the simplest common parent-forms—or: That is not the case, and the distinct species of organisms have originated independently of each other, and in that case can only have been created in a supernatural way, by a miracle. Natural evolution, or supernatural creation [2] of species—we must choose one of these two possibilities, for a third there is not.

But as Virchow, like many other opponents of the doctrine of evolution, constantly confounds this latter proposition with the doctrine of descent, and that again with Darwinism, it will not be superfluous to indicate here, in a few words, the limitation and subordination of these three great theories.

I. The general doctrine of development, the progenesis-theory or evolution-hypothesis (in the widest sense), as a comprehensive philosophical view of the universe, assumes that a vast, uniform, uninterrupted and eternal process of development obtains throughout all nature; and that all natural phenomena without exception, from the motions of the heavenly bodies and the fall of a rolling stone to the growth of plants and the consciousness of men, obey one and the same great law of causation; that all may be ultimately referred to the mechanics of atoms—the mechanical or mechanistic, homogeneous or monistic view of the universe; in one word, Monism.

II. The doctrine of derivation, or theory of descent, as a comprehensive theory of the natural origin of all organisms, assumes that all compound organisms are derived from simple ones, all many-celled animals and plants from single-celled ones, and these last from quite simple primary organisms—from monads. As [3] we see the organic species, the multiform varieties of animals and plants, vary under our eyes through adaptation, while the similarity of their internal structure is reasonably explicable only by inheritance from common parent-forms, we are forced to assume common parent-forms for at least the great main divisions of the animal and vegetable kingdoms, and for the classes, orders, and so forth. Thus the number of these will be very limited, and the primitive archigonian parent-forms can be nothing else than monads. Whether we finally assume a single common parent-form (the monophyletic hypothesis), or several (the polyphyletic hypothesis), is wholly immaterial to the essence of the theory of descent; and it is equally immaterial to its fundamental idea what mechanical causes are assumed for the transformation of the varieties. This assumption of a transformation or metamorphosis of species is, however, indispensable, and the theory of descent is very properly called also the "metamorphosis hypothesis," or "doctrine of transmutation;" as well as Lamarckism, after Jean Lamarck, who first founded it in 1809.

III. The doctrine of elimination, or the selection theory, as the doctrine especially of "choice of breed or selection," assumes that almost all, or at any rate most, organic species have originated by a process of selection; the artificial varieties under conditions of [4] domestication — as the races of domestic animals and cultivated plants — through artificial choice of breeds; and the natural varieties of animals and plants in their wild state by natural choice of breeds: in the first case, the will of man effects the selection to suit a purpose; in the second, it is effected in a purposeless way by the "struggle for existence." In both cases the transformation of the organic forms takes place through the reciprocal action of the laws of inheritance and of adaptation; in both cases it depends on the survival or selection of the better-qualified minority. This theory of elimination was first clearly recognised and appreciated in its full significance by Charles Darwin in 1859, and the selection-hypothesis which he founded on it is Darwinism properly so called.

The relation that these three great theories, which are frequently confounded, bear to one another may, according to the present position of science, be simply defined as follows: — I. Monism, the universal theory of development, or the monistic progenesis-hypothesis, is the one only scientific theory which affords a rational interpretation of the whole universe and satisfies the craving of our human reason for causality, by bringing all natural phenomena into a mechanical causal-connection as parts of a great uniform process of evolution. II. The theory of transmutation, or descent, is an essential and indispensable element in [5] the monistic development hypothesis, because it is the one only scientific theory which rationally explains the origin of organic species — that is to say, by transformation — and reduces it to mechanical principles. III. The theory of Selection or Darwinism is, up to the present time, the most important of the various theories which seek to explain the transformation of species by mechanical principles, but it is by no means the only one. If we assume that most species have originated through natural elimination, we also now know, on the other hand, that many forms distinguished as varieties are hybrids between two different varieties, and can be propagated as such; and it is equally well worthy of consideration that other causes are in activity in the formation of species of which, up to the present time, we have no conception. Thus it is left to the judgment of individual naturalists to decide what share is to be attributed to natural selection in the origin of species, and even at the present day authorities differ widely on the subject. Some give it a large share, and some a very small one in the result. Moritz Wagner, for instance, would substitute his own migration-hypothesis for Darwin's theory of selection; while I regard the action of migration, which acts as isolation or separation, as merely a special mode of selection. But these differing estimates of Darwinism are quite independent of the absolute [6] import of the doctrine of descent or of transformation, for the latter is as yet the only theory which rationally explains the origin of species. If we discard it, nothing remains but the irrational assumption of a miracle, a supernatural creation.

In this crucial and unavoidable dilemma, Virchow has declared himself publicly in favour of the latter, and against the former hypothesis. Every one who has attentively followed his occasional utterances on the theory of descent during the last decade with an unprejudiced eye and an unbiassed judgment, must be convinced that he fundamentally rejects it. Still, his dissent has always been so obscured, and his judgment on Darwinism in particular so wrapped in ambiguities, that an opportune conversion to the opposite side seemed not impossible; and many, even among those who stood near to Virchow—his friends and disciples—did not know to what point he was in fact an opponent of the evolution hypothesis in general. Virchow took the last step towards clearing up this matter at Munich; for after his Munich address there can be no farther doubt that he belongs to the most decided opponents of the whole theory of evolution, including those of inheritance and selection.

If any one still has doubts on the matter, let him read the jubilant hymns of triumph with which Virchow's friend and collaborator, Adolf Bastian, [7] greeted his Munich discourse. This "enfant terrible" of the school—this well-nicknamed "Acting privy counsellor of the board of confusion" [10]—whose merits in involuntarily advancing the cause of metamorphism I have already done justice to in the preface to the third edition of my "Natural History of Creation" [11]—expresses himself in the "Zeitschrift für Ethnologie," which is edited by him and Virchow (tenth yearly part, X. 1878, p. 66) as follows:—"At the Munich meeting of naturalists, Virchow by a few weighty words cleared the atmosphere, which was heavy and stifling under the pressure of the incubus called Descent, and once more freed science from that nightmare which it has so long—in many opinions so much too long—allowed to weigh upon it; freed it, let us hope, once and for ever. The forecasts of this storm were discernible many years since, and its whole course has been a strictly normal one. When the germs planted by Darwin, and that promised so much, were forced into growth by a feverish, hot-house heat, and began to sprout into sterile weeds, their small vitality was plain to our eyes. So long as the waves run too high under the pressure of a psychical storm, it is almost useless to protest against it, for [8] every ear is too much deafened by the noise all round to hear the voice of individuals. It is best to leave things to go their own way, deeper and deeper into the mire, till they come to a stand-still there of their own accord; for 'Quos deus vult perdere prius dementat.' Thus it is in this case. When the extravagances of the descent hypothesis, encouraged as they were by mutual incitement, had reached their highest pitch in the ravings that were uttered at Munich, the too pointed point broke in this superabundance of absurdity almost by its own pointedness, and so we were quit of it with one blow. Now, happily, all is over with the theory of descent, or ascent, but natural science will not on that account fare any the worse, for many of its adherents belong to her ablest youth, and as they now need no longer waste their best time on romantic schemes, they will have it to use at the orders and for the advancement of science, so as to enrich her through real and solid contributions."

Furthermore, Bastian quotes Virchow's maxim:—"The plan of organisation is immutable within the limits of the species; species is not produced from species." The fundamental teleological idea of that school, that each species has its constant and specific plan of structure, certainly cannot be more emphatically expressed. Thus it is undoubtedly certain that Virchow has become a Dualist, and [9] is as thoroughly penetrated by the truth of his principles as I, as a

Monist, am of mine. This is undoubtedly the upshot of his Munich address, though he is throughout careful to avoid acknowledging his chief standpoint in all its nakedness. On the contrary, even now he still veils his antagonism under the phrase, which is also a favourite with the clerical papers, that the theory of descent is an "unproved hypothesis." Now it is clear that this theory never will be "proved" if the proofs that already lie before us are not sufficient. How often has it been repeated that the scientific certainty of the hypothesis of descent is not grounded in this or that isolated experiment, but in the collective sum of biological phenomena; in the causal nexus of evolution. Then what are the new proofs of the theory of descent which Virchow demands of us? [10]

[10] "Wirkliche Geheime Ober-Confusionsrath."

[11] Translated under the supervision of E. Ray Lankester. London: C. Kegan Paul & Co.

CHAPTER II.

CERTAIN PROOFS OF THE DOCTRINE OF DESCENT.

All the common phenomena of Morphology and Physiology, of Chorology and Œkology, of Ontology and Paleontology, can be explained by the theory of descent, and referred to simple mechanical causes. It is precisely in this, viz., that the primary simple causes of all these complex aggregates of phenomena are common to them all, and that other mechanical causes for them are unthinkable — it is in this that, to us, the guarantee of their certainty consists. For this reason all these vast and manifold aggregates of facts are so many evidences of the doctrine of descent. This fundamental relation of facts has been so often expounded that I need dwell no farther on it in this place; those who wish for any closer discussion of it are referred to my "General Morphology" (vol. ii. chap. xix.), or "The History of Creation," [12] or "The Evolution of Man" (vol. i. p. 93). [13]

And where is yet farther proof of the truth of the [11] theory of descent to be found? Neither Virchow, nor any one of the clerical opponents and the dualistic philosophers who are perpetually reiterating this cry for more certain evidence, anywhere indicate where possibly such evidence is to be sought. Where in all the world can we discover "facts" which will speak more plainly or significantly for the truth of transmutation than the facts of comparative morphology and physiology; than the facts of the rudimentary organs and of embryonic development; than the facts revealed by fossils and the geographical distribution of organisms—in short, than the collective recognised facts of the most diverse provinces of biological science?

But I am in error—the certain proof that Virchow demands in order to be perfectly satisfied with the evidence, is to be supplied by "experiment, the test as well as the highest means of evidence." This demand, that the doctrine of descent should be grounded on experiment, is so perverse and shows such ignorance of the very essence of our theory, that though we have never been surprised at hearing it continually repeated by ignorant laymen, from the lips of a Virchow it has positively astounded us. What can in this case be proved by experiment, and what can experiment prove?

"The variability of species, the transformation of [12] species, the transition of a species into one or more new varieties," is the answer. Now, so far as these facts can be proved by experiment, they actually have long since been experimentally proved in the completest manner. For what are the numberless trials of artificial selection for breeding purposes which men have practised for thousand of years in breeding domestic animals and cultivated plants, but physiological experiments which prove the transformation of species? As an example we may refer to the different races of horses and pigeons. The swift race-horse and the heavy pack-horse, the graceful carriage-horse and the sturdy cart-horse, the huge dray-horse and the dwarfed pony—these and many other "races" are so different from each other, that if we had found them wild we should certainly have described them as quite different varieties of one species, or even representatives of different species. Undoubtedly, these so-called "races" and "sports" of the horse tribe differ from each other in a much greater degree than do the zebra, the quagga, the mountain horse, and the other wild varieties of the horse, which every zoologist distinguishes as "bonæ species." And yet all these artificial varieties, which man has designedly produced by selection, are descended from a single common parent-form, from one wild "true variety." The same is the case with the numerous and highly differing [13] varieties of pigeons. Domestic pigeons and carrier-pigeons, turbits and cropper-pigeons, fantail pigeons and owls, tumblers and pouters, trumpeters and laughing pigeons (or Indian doves), and the rest, are all, as Darwin has convincingly proved, descendants of a single wild variety, the rock-pigeon (*Columba livia*). And how wonderfully various they are, not only in general form, size, and colouring, but in the particular form of the skull, the beak, the feet, and so forth! They differ much more in every respect each from the others than the numerous wild varieties which, in systems of ornithology, are recognised as true varieties, and even as true species. It is the same with the different artificial varieties of apples, pears, pansies, dahlias, and so on; in short, of almost all the domestic varieties of animals and plants. We would lay particular stress on the fact that these artificial species which man has produced or created by artificial breeding and through experimental transformation out of one original species, differ far more one from another in physiological as well as in morphological conditions than the natural species in a wild state. With these it is self-evident that any proof by experiment of a common origin is wholly impossible. For, so soon as we subject any wild variety of animal or plant to such an experiment, we bring it under the conditions of artificial breeding. [14]

That the morphological conception of a Species is not a positive but only a relative conception, and that it has no other absolute or positive value than those other similar system-categories—sports, varieties, races, tribes, families, classes—is now acknowledged by every systematiser who forms an honest and unprejudiced judgment of the practical systematic distinction of species. From the very nature of the case there are no limits to arbitrary discretion in this department, and there are no two systematists who are at one in every instance; this one separating forms as true varieties which that one does not. (Compare on this point "History of Creation," vol. i., p. 273.) The conception of variety or species has a different value in every small or large department of systematic Zoology and Botany.

But the conception of species has just as little any fixed physiological value. In respect to this we must especially insist that the question of hybrid offspring, the last corner of refuge of all the defenders of the constancy of species, has at present lost all significance as bearing on the conception of species. For we know now, through numerous and reliable experiences and experiments, that two different true varieties can frequently unite and produce fertile hybrids (as the hare and rabbit, lion and tiger, many different kinds of the carp and trout tribes, of willows, brambles, [15] and others); and in the second place, the fact is equally certain that descendants of one and the same species which, according to the dogma of the old schools, could always effect a fertile union under certain circumstances, either cannot effect such a union or produce only barren hybrids (the Porto-Santo rabbit, the different races of horses, dogs, roses, hyacinths, &c.; see "History of Creation," vol. i., p. 146).

For a certain proof that the conception of species rests on a subjective abstraction and has a merely relative value—like the conception of genus, family, order, class, &c.—no class of animals is of so much importance as that of the Sponges. In it the fluctuating forms vary with such unexampled indefiniteness and variability as to make all distinction of species quite illusory. Oscar Schmidt has already pointed this out in the siliceous sponges and keratose sponges; and I, in my monograph, in three volumes, on the Calcareous Sponges (the result of five years of most accurate investigations of this small animal group), have pointed out that we may at pleasure distinguish 3, or 21, or 111, or 289, or 591 different species. I also believe that I have thus convincingly demonstrated how all these different forms of the calcareous sponges may quite naturally, and without any forcing, be traced to a single common parent-form, the simple—and not hypothetical, but existing at this present day—the simple Olynthus. [16] Hence I think I have here produced the most positive analytical evidence of the transformation of species, and of the unity of the derivation of all the species of a given group of animals, that is generally possible.

Properly, I might spare myself these disquisitions on the question of species, for Virchow does not go into this main question of the theory of descent—but this is very characteristic of his attitude. And just as he nowhere thoroughly discusses the doctrine of transformation, neither does he enter generally on the refutation of any of the other certain proofs of the doctrine of descent which we in fact possess at the present day. Neither the morphological nor the physiological arguments for the theory of descent, neither the rudimentary organs nor the embryonic forms, neither the paleontological nor the chronological argument are anywhere closely examined and tested as to their worth or their worthlessness as "certain proofs." On the contrary, Virchow takes them quite easily, sets them aside, and declares that "certain proofs" of the doctrine of descent do not exist, but remain to be discovered. To be sure, he does not indicate where they are to be sought, nor can he indicate it. How is this strange conduct to be explained? How is it possible that a distinguished naturalist should resist the most important step forward of modern natural science without in any way [17] specially investigating it, without even practically testing and refuting the most weighty arguments in its favour? To this question there is but one answer. Virchow is not generally intimate with the modern doctrine of evolution, and does not possess that knowledge of natural science which is indispensable for any well-grounded judgment on it.

After collecting and carefully reading all that Virchow, during many years, had written against evolution, I arrived at the conviction that he had not thoroughly read either Darwin's great work on the Origin of Species, nor any other work on the theory of descent, nor had he thought the matter out with such attention as so serious and intricate a subject absolutely demands. Virchow did with these works as it has been his well-known custom to do with many others—he hastily turned over the pages, caught at a few leading words, and without any farther trouble he has discoursed upon them, and, which is worst of all, has perpetuated these discourses through the press.

To excuse this conduct, and to account for Virchow's enigmatical position in the battle of evolution, we must consider what changes this highly-gifted and meritorious man has gone through in the course of the last thirty years. The most important and fruitful part of his life and labours was indisputably during [18] the eight years when he resided in Würzburg, from 1848 to 1856. There Virchow, with all the keenness of his youthful intellect, with a sacred enthusiasm for scientific truth, with indefatigable powers of work and the rarest insight, worked out that glorious reform of scientific medicine which will shine through all time as a star of the first magnitude in the history of medical science. In Würzburg, Virchow elaborated that comprehensive application of the cellular theory to pathology which culminates in the conception that the cell is an independent living elementary organism, and that our human organism, like that of all the higher animals, is merely a congeries of cells—a highly fertile conception, which Virchow now denies as resolutely as he then supported it. In Würzburg, twenty-five years since, I sat devoutly at his feet, and received from him with enthusiasm that clear and simple doctrine of the mechanics of all vital activity—a truly monistic doctrine, which Virchow now undoubtedly opposes where formerly he defended it. In Würzburg, finally, he wrote those incomparable critical and historical leading articles which are the ornament of the first ten yearly series of his "Archives" of pathological anatomy. All that Virchow effected as the great pioneer of reform in medicine, and by which he won imperishable honour [19] in the scientific treatment of disease,—all this was either carried out or preconceived in Würzburg; and even the celebrated "Cellular Pathology," a course of lectures which he delivered during the first year and a half after quitting Würzburg for Berlin, consists only of the collected and matured fruits of which the blossoms are due to Würzburg.

In the autumn of 1856 Virchow left Würzburg to settle in Berlin. The exchange of a narrow sphere of labours for a wider one, of small means and appliances for greater ones, proved unfavourable in this case, as in many similar cases. Since he has been in Berlin, in a "great Institution," and with luxurious appliances, all the scientific results which Virchow has as yet brought to light are not to be compared, either as to quality or quantity, to the grand and immortal achievements which he himself effected in the little institute of Würzburg with the scantiest means—a new proof of the maxim enunciated by me, and hitherto never confuted, that "the scientific results of an institute are in inverse proportion to its size." (See "The Aim and Methods of Modern Evolution." [14])

Still more grave is the circumstance that, since settling in Berlin, Virchow has more and more exchanged his theoretical scientific activity for practical [20] political life. It is well known how prominent a part he plays there in the Prussian Chamber of Representatives, how he raised himself to be the leader of the party of progress, and, to give this political position a broader basis, took part in the representation of the citizens of the capital; how he has taken a most active interest, as city commissioner, in all the petty anxieties and concerns which the charge of such a city as Berlin entails. I am far from blaming, as many have blamed, the political and civic activity to which Virchow has indefatigably devoted his best powers. If a man feels in himself the inclination and vocation with strength and talent enough, to play a conspicuous political part, by all means let him do so; but verily I do not envy him; for the satisfaction which is derived from the most successful and fruitful political activity is not, to my taste, to be compared with that pure and disinterested satisfaction of the mind which results from absorption in serious and difficult scientific labours. In the turmoil of the political and social struggle, even the most splendid civic crown will be dulled by the stifling dust of practical life, which never reaches the ethereal heights of pure science and never rests on the laurels of the thoughtful investigator. However, as I have said, that is a matter of taste. If Virchow really believes that he is doing a greater service to humanity by his practical [21] political life in Berlin than he formerly did by his theoretical scientific work in Würzburg, that is his affair; but for all that, in his former sphere he was incomparable, and cannot be replaced; in the latter this is not the case.

If a distinguished man, be he never so remarkable for uncommon power of work and universal gifts, passes the whole day in the friction of political party-struggles, and throws himself as well into all the petty and wearisome details of daily civic life, it is impossible for him to maintain the requisite feeling for the progress of science—particularly when it advances so rapidly and incessantly as is the case in our day. It is therefore quite intelligible that Virchow should soon have lost this feeling, and in the course of the last two decades have become more and more estranged from science. And this estrangement has at last led to so complete a change in his fundamental views, to such a metapsychosis, that the present Virchow of 1878 is hardly in a position to understand the youthful Virchow of 1848.

We have seen a similar mental change occur contemporaneously in our greatest naturalist, Carl Ernst von Baer. This gifted and profound thinker and biologist, whose name marks a new epoch in the history of evolution, had in his later years become wholly incompetent even to understand those most [22] important problems of his youthful labours which opened up new paths of inquiry. While in his early years he laid down principles of the greatest value to our modern doctrine of evolution, and even went very near to adopting this hypothesis into his system, at a later period he utterly denied it, and by his writings on Darwinism proved that he was no longer generally capable of mastering this difficult problem. As I am one of Von Baer's warmest admirers, and in my "Evolution of Man," as well as in the "History of Creation," and in other places, have most emphatically expressed that sincere esteem, I thought I might venture to forbear from calling attention to the discrepancy between the lucid, monistic principles of Von Baer in his youth, and the confused dualistic views of his old age. But as many opponents of Darwinism—and among them particularly the Old Catholic philosopher of Munich, Huber, who has written a series of articles in the "Augsburger Zeitung"—have made constant capital out of the harmless talk of the feeble old Von Baer, I must in this place explicitly declare that this dualistic prating of the old man is quite incapable of shaking the monistic principles of the young and enterprising pioneers of science, or of giving them the lie.

In his autobiography Von Baer gives us the explanation of this striking contradiction. In 1834 he [23] entirely and for ever abandoned the province of the history of development, at which for twenty years he had laboured incessantly, and where he had earned splendid laurels. To escape from the haunting and importunate ideas of the science which had so wholly absorbed him, he fled from Königsberg to Petersburg, and subsequently busied himself in scientific inquiries of a quite different character. Twenty-five long years passed by, and when Darwin's work appeared in 1859, Von Baer had too long undergone a metapsychosis to be able to understand it. In Von Baer, as in Virchow, the course of this remarkable metapsychosis is highly instructive, and will itself afford to the thoughtful psychologist an interesting evidence of the doctrine of evolution.

However, the lack of comprehension of our modern evolution-hypothesis is easier to explain in Virchow's case than in Von Baer's, for this reason: morphological knowledge was greatly lacking to Virchow, while Von Baer possessed it in the highest degree. Now morphology is precisely that very department of inquiry in which our theory of descent has its deepest and strongest roots, and has matured the most glorious fruits of knowledge. The study of organic forms, or morphology, is thus, more than any other science, interested in the doctrine of descent, because through this doctrine it first obtained a practical knowledge [24] of effective causes, and was able to raise itself from the humble rank of a descriptive study of *forms* to the high position of an analytical science of *form*. It is true that by the beginning of this century the most comprehensive branch of morphology—*i.e.,* comparative anatomy—which was founded by Cuvier and splendidly developed by Johannes Müller, had laid the foundations on which to build a truly philosophical science of form. The enormous mass of various empirical material, which had been accumulated by descriptive systematists and by the dissections of zootomists since the time of Linnæus and Pallas, had already been abundantly matured and utilised in many ways for philosophic purposes by the synthetic principles of comparative anatomy. But even the most important universal laws of organisation—of which the old system of comparative anatomy was one—had to take refuge in mystical ideas of a plan of structure and of creative final causes (*causæ finales*); they were incapable of arriving at a true and clear perception of effective mechanical causes (*causæ efficientes*). This last, most difficult, and grandest problem, Charles Darwin was the first to solve in 1859, by setting Lamarck's theory of descent, which was already fifty years old, on a firm footing by his own theory of selection. By this hypothesis it was first made possible to fit together the rich materials which had [25] been previously amassed, into the splendid edifice of the mechanical science of form. (See my "General Morphology," vol. i. chap. iv.)

The immeasurable step which Darwin thus made in organic morphology can be adequately appreciated only by those who, like myself, were brought up in the school of the old teleological morphology, and whose eyes were suddenly opened by the theory of selection to a comprehension of that greatest of all biological riddles, the creation of specific forms. The dogma of creation, the mystic and dualistic doctrine of the isolated creation of each separate variety, was annihilated at one blow; the belief in transmutation has now for ever taken its place—the mechanistic and monistic doctrine of the metamorphosis of organic forms, of the descent of all the species of one natural class from a common parent-form. How complete a change the science of mechanical morphology has by this means been compelled to undergo, I have endeavoured to point out in my "General Morphology;" and any one who wishes to convince himself clearly of what an enormous revolution has been brought about, particularly in comparative anatomy, may compare the "Outlines of Comparative Anatomy" (Grundzüge der vergleichenden Anatomie), by Carl Gegenbaur, 1870, and the latest edition of his "Elements" (Grundrisses), with the old text-books of that science. [26]

Virchow has no suspicion even of all these immeasurable strides in morphology, for this department always lay out of his ken. His great reforms in pathology were founded in the province of physiology, and more especially in cellular physiology. But within the last twenty years these two main branches of biological inquiry have grown more and more apart. The great Johannes Müller was the last biologist who was able to keep these departments of organic inquiry together, and who won equally immortal honours in both divisions of the subject. After Müller's death in 1858 they fell asunder. Physiology, as the science especially of the functions or living activity of the organism, addressed itself more and more to exact and experimental methods: morphology, on the contrary, as the science of the forms and structure of animals and plants, could naturally make but very small use of this method; it must take refuge more and more in the history of evolution, and so constitute an historical natural science. It was on this very historical and genetic method of morphology, in contradistinction to the exact and experimental method of physiology, that I based my Munich address; and if Virchow in his answer had really and thoroughly refuted this position, instead of fighting with mere phrases and denunciations, this radical opposition would have been well worthy of the fullest discussion. At the same time [27] I have no wish to reproach Virchow for being wholly fettered by the one-sided views of the modern school-physiology, nor because morphology lies so far out of his ken that he has not been able to form an independent judgment of its aims and methods; but when, in spite of all this, he on every occasion lets fall a disparaging judgment of it, we must dispute his competence. It is true that in his Munich address he emphasises the statement, "That which graces me best is that I know my ignorance," by printing it in italics. I only regret that I am forced to deny his possession of this very grace. Virchow does not know how ignorant he is of morphology, else he would never have uttered his annihilating verdict on it, else he would not continually designate the study of the theory of descent as dilettanteism and vain dreaming, as "a fanciful private speculation which is now making its way in several departments of natural science." In truth, Virchow does me greatly too much honour when he designates as my "personal crotchet" an idea which for the last ten years has been the most precious common possession of all morphological science. If Virchow were not so unfamiliar with the literature of morphology, he must have known that it is penetrated throughout by this principle of descent, that every morphological inquiry which conscientiously pursues a well-considered problem now assumes the [28] doctrine of descent as granted and indisputable. Of all this he is ignorant, and so it is intelligible that he should continue to demand "certain proofs" of this hypothesis, although those proofs have long since been produced.

[12] Vol. ii., p. 334 of translation.

[13] London: C. Kegan Paul & Co. 1879.

[14] Jena, Zeitschriften für Naturwissenschaft, 1875. Vol. x. Supplement.

[29]

CHAPTER III.

THE SKULL THEORY AND THE APE THEORY.

Inasmuch as Virchow persists in treating the theory of descent as an "unproved hypothesis," inasmuch as he ignores all the forcible evidences of that hypothesis, he deprives himself of the right of speaking a decisive word in this, the most important scientific dispute of the present day. Virchow is, in fact, simply incompetent in the great question of evolution, as he is deficient in the greater part of that knowledge—more especially morphological knowledge—which is indispensable to forming a judgment upon it. Hence on the turning-point of the whole matter—viz., the problem as to the origin of species—he can have no opinion, as he has never turned his attention to the systematic treatment of species: those transitions of one species into another, which he asks to see, abound on all sides, as is well known to every systematic naturalist. Only consider, for example, the genera of Rubus and Salix among the living plants of the present period, and the Ammonites and Brachiopoda among [30] extinct animals. Hence, too, Virchow can have no independent views as to the historical development of the higher from the lower animals, because the abundant living forms of the lower animals are almost unknown to him, and because he has hardly any conception of the marvellous strides which hundreds of industrious workers have made in this very department within the last twenty years. But there can be no doubt, indeed it is already universally acknowledged, that it is precisely the comparative anatomy of the lower—nay, of the very lowest animals—that has solved the greatest riddles of life, and removed the greatest obstacles from the path of the doctrine of descent. He simply ignores the fact that true Monads actually exist, and have been positively identified by many different observers as structureless "organisms without organs," and he turns out the poor Bathybius with a kick. And yet I believe that in "Kosmos" [15] I have conclusively proved that Monads must retain their vast elementary importance whether the Bathybius actually exists or not.

But even as regards the higher animals—nay, even as to the comparative anatomy of the highest next to man, the apes—Virchow stands apart, not understanding the views of modern morphology.

We must here examine more closely into this, [31] because it is precisely in this department that Virchow's only morphological experiments have been made; viz., his investigations as to the skulls of apes and of men. This is precisely the one only point on which he has sought a closer acquaintance with morphology, and precisely here it is most clearly to be seen how little he is acquainted with the recent advances our science has made, and that he has hardly any conception of the extraordinary importance to that science of the theory of descent.

The skull theory, as is well known, has for a long time been a very favourite theme, not only with prominent naturalists, but also with talented amateurs. Undoubtedly the skull, viewed as the bony capsule which encloses our most important organ of sense, our brain, has a special claim to morphological importance; for the general conformation of the skull corresponds on the whole to the development of the brain, and its inner surface gives an approximate idea of the outer surface of the brain. In this correspondence lies the only sound kernel of the sickly, overgrown fancies of phrenology. The various development of the skull allows of an approximate inference as to the various degrees of development of the brain and of the mental faculties. The comparative study of the skulls of the vertebrate animals had excited the lively interest of morphologists by the end of the last [32] century, when comparative anatomy was beginning to constitute a special science; and the genetic inquiry as to the morphological significance and development of the skull soon grew out of it. It was no less a man than our greatest German poet who first answered this question, and propounded the theory that the skull was neither more nor less than the modified foremost end of the vertebral column, and that the separate groups of bones which lie behind one another in the human skull, as in that of all the higher vertebrata, answer to the separate modified vertebræ. This "vertebral theory" of the skull, which Von Goethe and Oken simultaneously and independently attempted to prove, aroused universal interest and maintained its ground for seventy years, while many attempts were made to improve and enlarge upon it in detail.

A quite new light was thrown on this, as on every other morphological question, as soon as Darwin in 1859 had once more put into our hands the torch of the doctrine of descent. The inquiry as to the origin of the skull now assumed a real and tangible form. Since all vertebrate animals, from fishes up to man, agree so completely as to their essential internal structure that they can be rationally conceived of no otherwise than as branches of one stock and as descendants of one parent-form, the distinctly formulated question as to [33] the skull theory which now started into prominence was this: "How, historically, has the skull of man and of the higher animals originated from that of the lower animals? How is the development of the bones of the skull from the vertebræ to be proved?" The answer to these difficult questions was supplied by the first comparative anatomist of the present day, by Carl Gegenbaur. After Huxley had pointed out that the ontogenesis or individual development of the skull by no means favoured the older hypothesis of Goethe and Oken, Gegenbaur brought forward evidence that the fundamental idea of that theory was correct; that the skull does in fact correspond to a series of coalescent vertebræ, but that the separate bones of the skull are not to be regarded as representing parts of such modified vertebræ. The skull-bones of all recent vertebrate animals are rather, for the most part, dermal bones, which have come into closer connection as supplementary to the cartilaginous primitive skull. We can even now trace the number and position of the original vertebræ, from which this primitive skull originated, by the number of the vertebral arches (gill-arches) which are attached to it, as well as by the number and position of those vertebræ, from nine to ten. Of all the recent vertebrata, the cartilaginous fishes, or Selachians, have most nearly preserved the form and structure of this primordial skull. These Selachians, the Rays and [34] Sharks, are on the whole the creatures which throw the clearest light on the history of the lineage of the vertebrata and on the organisation of our primeval fish-natured ancestors. It is one of the particular merits of Gegenbaur that he clearly and firmly established the place in nature of the Selachians as the common ancestors of all vertebrate animals from fish up to man.

None but those who have thoroughly studied the comparative morphology of the vertebrata, who have sought the genetic issue from that labyrinth of intricate morphological problems at the hands of the theory of descent, can duly value the immeasurable service which Gegenbaur has done by this and other "Investigations into the Comparative Anatomy of the Vertebrata." These investigations are as much distinguished by a profound knowledge and careful working out of the wonderfully-extensive empirical materials for the subject, as by their critical acumen and philosophic grasp. At the same time they set in the clearest light the immeasurable value of the theory of descent in the causal explanation of the most difficult morphological problems. Gegenbaur might, therefore, with perfect right, enunciate this axiom in the Introduction to his "Comparative Anatomy." "The theory of descent will at once find a touchstone of proof in comparative anatomy. Up to this time no experience in comparative anatomy has [35] transpired which contradicts that theory; on the contrary, they all lead up to it. Thus it will receive back from science that which it has given to scientific method: clearness and certainty." In point of fact we can adduce no morphological investigations which better support this declaration than those very phylogenetic researches "as to the cranium of the Selachians, as a basis for the critical examination of the genesis of the cranium of the vertebrata," 1872. As Virchow had formerly thoroughly studied the old skull-hypothesis, and in his admirable discourse on "Goethe as a Naturalist," 1861, had given an excellent exposition of it; as moreover he had produced most valuable contributions to the normal and pathological anatomy of the human skull, we might have expected that he would have received Gegenbaur's grand reform of the theory of the skull, and historical solution of the skull-problem, with the greatest interest, and have made it the clue to his own further researches. But we seek in vain through Virchow's latest contributions to the study of the human skull, for any indication of his knowing or appreciating Gegenbaur's investigations. On the contrary, we see him persistently moving, without any clear goal in view, on that trodden and devious path of investigation which finds the highest aim of craniological science in the measuring of skulls, or craniometry. [36]

We are far from undervaluing the full significance of the results of exact and careful descriptions and measurements of various conformations of the skull as an empirical basis for a true and scientific study of the skull—*i.e.*, for comparative and genetic craniology. But still we must say that the way and method by which this skull measurement has, for ten years now, been pursued by numerous craniologists can never yield corresponding scientific results; on the contrary, though it is cried up as the "exact morphology" of the skull, it simply loses itself in the domains of harmless trifling. A large amount of time has in the last ten years been squandered in disputes as to the best method of measuring skulls, while the craniologists concerned have not, in the first place, answered the obviously most important question: What end they propose to gain by this specialist measuring, what proposition they mean to prove by it? Most of those numerous skull measurers know nothing beyond the perfect human skull, or at most the skulls of a few other mammalia, while the comparative morphology and historical development of the crania of the lower vertebrata are wholly unknown to them; and yet these last contain the true key to the comprehension of the others. One single month devoted by these "exact skull measurers" to the study of Gegenbaur's theory of the skull, and to testing the hypothesis by the skulls of [37] Selachians, would have yielded them more fruit and have given them more light than long years of describing and measuring human skulls, however various.

Virchow himself affords the most striking example of the usual results of this so-called "exact method" of studying skulls. In his popular essay on "The Skulls of Men and Apes," 1870, he concludes with this notable proposition: — "It is therefore self-evident that Man can never by any progressive development have originated from the Apes." Every evolutionist who is familiar with the surprising facts of comparative morphology will draw from them the opposite conclusion: "It is self-evident that Man could only have originated from the progressive development of the Ape (organism)."

This brings us to that question which, in the popular treatment of the theory of descent, is justly considered as its most important outcome and as the keystone of the evolutionist edifice — to the well-known proposition, "Man is descended from the Ape." While we simply ignore all the misrepresentation, distortion, and misinterpretation which this ape, or pithecoid hypothesis, has met with on all sides, we will only remark that this fundamental proposition, in the sense of our modern doctrine of evolution, can rationally have only this plain meaning: that the [38] human species as a whole was long since developed from the order of apes, indeed actually from one (or perhaps more) long since extinct form of ape; the immediate progenitors of man in the long series of his vertebrate ancestry were apes or ape-like animals. Of course none of the now surviving species of apes is to be regarded as the unaltered posterity of that primeval parent-form. Virchow, however, understanding the "ape question" in this sense, answers it, as Bastian also does, with the most positive contradiction. "We cannot teach the doctrine that man is descended from apes or from any other animal, for we cannot regard it as a real acquisition of science" (p. 31). Although I myself, in direct opposition to this view, and in agreement with almost all my professional colleagues, look upon the descent of man from apes as one of the surest of phylogenetic hypotheses, I will here expressly admit that the *relative* certainty of this, as of all other historical hypotheses of descent, is not comparable with the *absolute* certainty of the general theory of descent. It is now ten years since I first explicitly stated (in my "Natural History of Creation," vol. ii. p. 358): "The pedigree of the human race, like that of every animal or plant, remains in detail a more or less approximate general hypothesis. This, however, in no way affects the application of the theory of descent [39] to man. In this, as in all researches into the derivation of our organism, we must distinguish between the *general theory* of descent and the *specific hypothesis* of descent. The general theory of descent claims full and permanent value, because it is inductively based on the whole range of common biological phenomena and on their internal causal connection. Each special hypothesis of descent, on the other hand, is conditional as to its specific value on the existing state of our biological information, and on the extent of those objective empirical grounds on which we deductively found the hypothesis, by our subjective inferences." And I must here emphatically add that I have on every opportunity repeated that reservation, and have always insisted on the difference which exists between the absolute certainty of transmutation in general and the relative certainty of each individual specific pedigree. So that when Semper and others of my opponents assert that I teach my specific genealogies as "infallible dogmas," it is simply false. I have, on the contrary, pointed out on all occasions that I regard them only as *heuristic or provisional hypotheses*, and as a means of investigating the actual relations of cognate races of organic forms more and more approximately.

Since the conception of the natural animal system as a hypothetical genealogical tree, and the [40] phylogenetic interpretation of morphological affinity which that conception involves, afford in fact the only rational interpretation of that affinity in general, my first genealogical attempts soon found many imitators, and at the present time numerous industrious labourers in the different departments of systematic zoology are endeavouring to find in the construction of such hypothetical genealogies the shortest and completest expression of the modern conception of structural affinity. If Virchow had not been as ignorant of the true significance and method of systematic morphology as he is of its progress and scientific contents, he must certainly have known this, and then he would surely have withheld his mockery of all these grave phylogenetic studies as "personal crotchets" and worthless dreams.

What mighty strides towards a mechanical morphology we have made by this phylogenetic working out of the system, and how much light and life it has at once thrown into the system that before was dead and cold, can only be known to those who have long and deeply studied specific systematisation and the grouping of species; Virchow has not the remotest suspicion of it. Moreover, these attempts have now proceeded so far, that a large proportion of the phylogenetic hypotheses are regarded as very nearly certain, and can hardly [41] undergo any further essential modifications; while the greater number of them are still in an unfixed state, and one systematist tries to improve them in this direction, and another in that.

The following phylogenetic hypotheses are held to be almost certain : — — The descent of many-celled animals from single-celled, of the Medusæ from the hydroid Polyps, of the jointed from the unjointed worms, of the sucking from the gnawing insects, of amphibious animals from fishes, of birds from reptiles, of the placental mammalia from the marsupials, and so forth. I personally consider the descent of man from the apes as equally certain; nay, I regard this most important and pregnant genealogical hypothesis as one of those which, up to the present time, rest on the best empirical basis.

Huxley, in particular, fifteen years ago, in his celebrated "Man's Place in Nature," 1863, so admirably proved the undoubted "descent of man from apes," and so clearly discussed all the relations that had to be taken into consideration, that very little was left to others to do. The result of his comparative morphological investigations is contained in this proposition — — " If we take up a system of organs, be it which we will, the comparison of its modifications throughout the series of apes leads us to the same conclusion: that in every single visible [42] character man differs less from the higher apes than these do from the lower members of the same order." It is therefore impossible for any objective zoologist, according to the principles of comparative systematisation, to ascribe to man any other place in the animal world than in the order of apes; and it is quite immaterial whether we designate this individual group as the Order of Apes, or, with Linnæus, as the Primates. For the phylogenetic construction of the system, the common descent of man and of apes from one common parent-form, necessarily follows from this inevitable grouping, and on this proposition only all the general inferences of the "ape-hypothesis" depend. As to what that common parent-form of men and apes may have been, very different views might probably be brought on opposite sides; but any one who knows the collected facts that bear upon the matter, and estimates them impartially, must, in conclusion, arrive at the certain conviction that that hypothetical and long-since extinct parent-form can only have been genuine apes; that is to say, of the placental mammalian type, such as when we see them now living before our eyes we unhesitatingly class, on the ground of their zoological characters, as true apes, in the order of Apes or Primates.

In this, and all other sound phylogenetic hypotheses, we may most easily attain to a conviction of their [43] truth by taking into consideration and comparison the other possible hypotheses. But in fact no single opponent of the ape-hypothesis has been able to combat it with any other phylogenetic hypothesis that has the faintest glimmer of probability. Not one opponent has suggested, or can suggest, any other animal form that can serve as our nearest ancestor than the ape. No one has ever reproached me by saying that Mother Nature has endowed me with too little imagination; on the contrary, I am often accused of having a superfluity of that gift of the gods; but I have often and repeatedly exerted my imagination to picture to myself any known or unknown animal-form as the nearest parent-form to man in the place of the apes, and have always found myself under the necessity of falling back upon the stock of apes. Let me conceive of the outward conformation and the internal structure of the nearest mammalian ancestors of men as I will, I am always forced to acknowledge that this hypothetical parent-form ranges under the zoologically-conceived order of apes, and cannot possibly be separated from the Simiadœ or Primates. If, in spite of this, any one chooses, out of a "personal crotchet," to accept some other series of unknown animal ancestors of man that have nothing to do with apes, that is but a mere empty hypothesis floating in the air. Our ape-hypothesis, on the other hand, is objectively and [44] thoroughly proved by the essential agreement of the internal bodily structure of man and of apes, and by the identity of their embryonic development, as I have fully shown in my "Evolution of Man" (chaps. xix. and xxvi.) The mode and manner in which he here puts palæontology in the foreground, and throws on the theory of descent the task of producing an unbroken gradation of fossil transitional forms between the apes and man, is very indicative of Virchow's ignorance of this zoological question—in which I, as a professional zoologist, must decisively declare his incompetence. The reasons why such a solution of the problem is not to be expected, the extraordinary imperfection of the palæontological record, the natural impediments to the palæontological evidence of the genealogical table, have been so lucidly unfolded by Darwin himself (chaps. ix. and x. of the "Origin of Species") that I am obliged once more to come to the conclusion that Virchow has never read it with any attention.

Besides, long before Darwin, the gifted Lyell, the great originator of modern geology, showed clearly and convincingly how, for many reasons, the greater part of the fossil series must remain most imperfect, and these reasons were at a later period so often and so fully discussed (by myself among others, in chap. xv. [45] of the "History of Creation," vol. ii. pp. 24-32) that it is wholly superfluous once more and in this place to state these well-known and time-worn questions. It only shows how little Virchow was acquainted with geology and palæontology, and what a limited judgment he can form of these historical causal relations.

[15] Vol. i. p. 293.

[46]

CHAPTER IV.

THE CELL-SOUL AND CELLULAR PSYCHOLOGY.

No attack in Virchow's Munich address surprised me so much, and none so plainly betrayed the subversion of his most important scientific views, as that which he directed against my observations on psychology and cellular physiology. A mystic dualism in his fundamental views is here revealed, which stands in the sharpest contrast to the mechanical monism formerly upheld by the famous pathologist of Würzburg.

In my Munich discourse (p. 12), I had alluded to the "grand and fruitful application which Virchow had made, in his system of cellular pathology, of the cell-theory to the general province of theoretic medicine;" and as a logical amplification of that idea, I asserted emphatically that we must ascribe an independent soul-life to every individual organic cell. "This conception is validly proved by the study of infusoria, amœbæ, and other one-celled organisms; for, in these individual, isolated, living cells we find the same manifestations of soul-life—feelings, and ideas [47] (mental images), will and motion, as is in the higher animals compounded of many cells" (p. 13). Virchow now rises up in the strongest protest against this theory of a cellular sensibility, which I regard as the inevitable consequence of his early views of cellular physiology; it is to him "mere trifling with words." He combats with equal decisiveness "the scientific necessity of extending the province of psychical processes beyond the circle of those bodies in and by which we actually see them exhibited." He further says, "If I explain attraction and repulsion as psychical phenomena, I simply throw the psyche out of the window; the psyche ceases to be a psyche." Finally he says, "I assert without any hesitation that for us the sum total of psychical phenomena is connected with certain animals only, and not with the collective mass of all organic beings; nay, not even with all animals in general. We have no ground as yet for speaking of the lowest animals as possessing psychical properties; we find such properties only in the higher grades, and with perfect certainty only in the very highest."

When I first read this and other astounding statements in Virchow's paper, I involuntarily asked myself, "Can this be the same Virchow from whom, twenty-five years ago, I learnt in Würzburg that the soul-functions of man and animals depend on mechanical [48] processes in the soul-organs; that these organs are, like all other organs, composed of cells, and that the functional activity of an organ is nothing more than the sum of the activity of all the cells which compose it? Is this the same Virchow whose most vital doctrine it was that all the physical and psychical processes of the human organism were to be referred to the mechanics of cell life; who supported the view of the unity of all the phenomena of life with the same emphasis with which we are now obliged to defend it against his attacks?"

In fact, and beyond a doubt, we have here a new proof of Virchow's complete change in all fundamental scientific principles. For the cellular psychology which I advance is only a necessary consequence of the cellular physiology promulgated by Virchow. His present opposition to the former is either a renunciation of the latter or an untenable and inconsequent position. To explain this astonishing metapsychosis, we shall do well first to glance at the soul in general, and then give particular consideration to the cell-soul.

What is the Soul or Psyche? The innumerable different answers which have been given to this crowning question of psychology, may collectively, when freed from all extraneous matter, be brought under two groups which we may shortly designate as the dualistic and the monistic soul-hypothesis. According [49] to the monistic (or realistic) soul-hypothesis, the "soul" is nothing more than the sum or aggregate of a multitude of special cell-activities, among which sensation and volition—sensual perception and voluntary movement—are the most important, the most common, and the most widely diffused; associated with these in the higher animals and in man, we find the more developed activities of the ganglionic cells which are included under the conceptions of Thought, Consciousness, Intellect, and Reason. Like all the other functional-activities of the organic cells, these soul-functions depend ultimately on material phenomena of motion, and more particularly on the motions of the plasson-molecules or plastidules, the ultimate atoms of the protoplasma, and perhaps of the nucleus also; therefore we should be able actually to grasp and explain them, as well as every other cognisable natural process, if we were in a position to refer them to the mechanics of atoms. This monistic soul-hypothesis, then, is at bottom mechanistic. If psychical mechanics—psychophysics—were not so infinitely complex and involved, if we were in a position to take a complete view of the historical evolution of the psychic functions, we could reduce the whole of them (including consciousness) to a mathematical "soul-formula."

According to the opposite, or dualistic (or spiritualistic) soul-hypothesis, the soul is, on the contrary, a [50] peculiar substance, which most people somewhat grossly conceive of as a gaseous body, while others picture it with more subtlety, as an immaterial essence. This "soul-substance" subsists independently of the animal-body, and stands in only a temporary connection with certain organs of that body—the soul-, or mental-organs. It has been imagined that this soul-matter, which resembles that imponderable ether which is the medium of light, is diffused between the ponderable molecules of the soul-organs and especially of the nerve-cells, and that this connection of the imponderable "soul" with the ponderable body subsists only so long as the individual life lasts. At the instant of the first beginning of the individual organism, at the moment of generation, this imponderable "soul" passes into the body, and at the instant of death, at the annihilation of the living individual, it again quits the body. This mystical or dualistic soul-hypothesis, which, as is well known, is to this day universally accepted, is fundamentally vitalistic, inasmuch as it regards the force which is bound up with the soul-substance, like the "vital force" of a past time, as a peculiar force quite independent of mechanical forces. This force does not depend on the material phenomena of motion, and is quite independent of the mechanics of atoms. The highest law of modern natural science, the law of [51] the conservation of force, has, therefore, no application in the region of soul-life, and that mechanical causality which prevails throughout all the processes of nature does not exist for the soul. The Psyche, in a word, is a supernatural phenomenon, and the supernatural department of the spiritual world stands free and independent of the natural department of the material world.

If we now compare the psychological views of the youthful and unprejudiced Virchow of Würzburg with those of the older and mystical Virchow of Berlin, there can be no doubt in the minds of the impartial that the former, a quarter of a century ago, was as decided and logical a monist as the latter is at present a confessed and convicted dualist. The distinguished position which Virchow, twenty-five years since, won by his natural conception of the nature of man, and the great fame which he then earned in the fight for the truth, rest precisely on this, that on every occasion he maintained with his utmost vigour the unity of all vital phenomena, and asserted their mechanical character. All organic life, even the soul-life, rests on mechanical principles, on that causal mechanism of which Kant said that "it alone contained a practical interpretation of nature," and that "without it no natural science can exist." On this point Virchow says well in his discourse on "Efforts at Unity in Scientific Medicine," [52] 1849: — "Life is only a peculiar sort of mechanics, though it is indeed the most complex form of mechanics; that in which the usual mechanical laws fall under the most unusual and manifold conditions. Thus life, compared with the universal processes of motion in nature, is a thing peculiar in itself; but it does not constitute a diametrical, dualistic opposition to those laws; it is only a peculiar species of motion. The motion itself is a mechanical one, for how should we become cognisant of it if it were not based on the sensible properties of bodies? The media of the motion are certain chemical matters, for we recognise none but chemical matter in bodies. The individual acts of motion reduce themselves to mechanical, or physico-chemical, modifications of the constituent elements of the organic unities, the cells and their equivalents." These and many similar utterances in Virchow's earlier writings, and especially in the essay I have mentioned, "On the Mechanical Conception of Life," leave no doubt that he formerly supported, with a clear conscience and his utmost energy, in psychology as in the other collected departments of physiology, that very mechanical standpoint which we to-day accept as the essential basis of our monism, and which stands in irreconcilable antagonism to the dualism of the vitalistic doctrine. To none of my teachers am I so deeply indebted for my emancipation from all the [53] prejudices of the dualistic doctrine, and for my conversion to the monistic, as to Rudolf Virchow; for it was his superior guidance which most firmly convinced me, and many others, of the exclusive importance of the mechanical view of nature. He led me to a clear recognition of the fact that the nature of man, like every other organism, can only be rightly understood as a united whole, that this spiritual and corporeal being are inseparable, and that the phenomena of the soul-life depend, like all other vital phenomena, on material motion only — on mechanical (or physico-chemical) modifications of cells. And it was in perfect agreement with my most honoured master that I subscribed then, and at this day still subscribe, to the proposition with which he, in

September 1849, closed the preface to the above-mentioned "Efforts at Unity." "It is possible that I may have erred in details; in the future I shall be ready and willing to acknowledge my mistakes and to rectify them, but I enjoy this conviction, that I shall never find myself in the position of denying the principle of the unity of the human nature with all its consequences!"

To err is human! Who can say to what diametrical contradiction to his firmest convictions man may not in the future be driven by his adaptation to new relations in life? If we compare these stout monistic declarations of 1849 and 1858 with the equally decided dualistic [54] utterances in Virchow's Munich address of 1877, we perceive that he could not give the lie more fiercely to his former fundamental opinions than he has there done. Not quite twenty years have passed by, and yet, in the course of that time, in Virchow's views of the universe, in his conception of human nature, and of the soul-life, a change has been effected than which we can conceive of no greater. We learn to our surprise that psychical and corporeal processes are wholly different phenomena; that no scientific necessity whatever exists for extending the province of psychical processes beyond the circle of those bodies in which, and by which, we see them actually exhibited. "We may ultimately explain the processes of the human mind as chemical, but at any rate, it is not yet our business to amalgamate these two subjects!"

From the whole psychological discussion which is involved in Virchow's Munich address, it is made clear that at the present time he regards the "soul" in a purely dualistic sense as a substance, an immaterial essence which only temporarily takes up its abode in the body. Highly characteristic of this is the remarkable sentence, "If I explain attraction and repulsion as psychical phenomena, I simply throw the psyche out of the window; the psyche ceases to be a psyche." If we substitute for the word "psyche" the word which corresponds to Virchow's earlier mechanistic [55] view—the word "motion" (or peculiar mode of motion)—the sentence runs thus: "If I explain attraction and repulsion as phenomena of motion, I simply throw motion out of the window."

Almost more remarkable is Virchow's assertion that the lowest animals have no psychic properties; that, on the contrary, "these are only to be found in the higher, and, with perfect certainty, only in the highest animals." It is only to be regretted that Virchow has not here stated what he understands by the higher and the highest animals; where that remarkable dividing line is, beyond which the soul suddenly appears in the hitherto soulless body. Every zoologist who is in some degree familiar with the results of comparative morphology and physiology will here clasp his hands in astonishment, for by this proposition Virchow seems to mean that we must ascribe a soul-life only to those animals in which special soul-organs, in the form of a central and peripheral nerve-system, are developed from sense-organs and muscles. But it is admitted that all these different soul-organs with their characteristic properties have originated from single cells through the division of labour (differentiation); and the nerves and muscles especially have been developed by differentiation from the neuro-muscular cells. The cells from which all these different nerve-cells, muscle-cells, mind-cells, and so forth, are derived, are [56] originally the simple neutral cells of the epithelium of the ectoderm or exterior germ-layer, and these cells, again, like all the cells of many-celled animal bodies, originated in the repeated division of one single original cell, the ovum-cell.

The individual development or ontogenesis of each of these many-celled animal-forms, brings this histological process of development so clearly and evidently before our eyes that we can but directly infer from it the truth of the phylogenesis, or gradual historical evolution of the soul-organs. The association of cells and the division of labour among them are the modes by which, in the first instance, the compound many-celled organism has originated, historically, from the simple one-celled organism. And an impartial comparative consideration teaches us in the clearest way that a functional-activity of the soul-cells exists in the lowest one-celled animals as well as in the highest and many-celled; in the infusoria as well as in man. Volition and sensation, the universal and unmistakable signs of soul-life, may be observed among the former as well as in the latter. Voluntary motion and conscious sensation (of pressure, light, warmth, &c.) come under our observation so undoubtedly in the commonest forms of infusorial animals—for instance the Ciliata, that one of their most persevering observers, Ehrenberg, asserted undeviatingly to the day of his death that all Infusoria [57] must possess nerves and muscles, organs of sense and of soul, as well as the higher animals.

It is well known that the enormous advance which our science has lately made in the natural history of these lowest organisms culminates in the statement—clearly made by Siebold thirty years since, but only recently "ascertained as proved"—that these minute creatures are *one-celled*, and that in the case of these infusoria one single cell is capable of all the various vital functions—including soul-functions—which in the zoophytes (plant-animals), as the hydra and the sponges, are distributed among the cells of the two germ-layers, and in all the higher animals among the different tissues, organs, and apparatus of a highly developed and constructed organism. The psychic functions of sensation and voluntary motion, which are here distributed to such very various organs and tissues, are in the infusoria fulfilled by the neutral plasson material of the cell, by the protoplasma, and possibly also by the nucleus (compare my treatise "The Morphology of the Infusoria." Jena, Zeitschriften, 1873, vol. vii. p. 516). And just as we must attribute to these primary animal forms an independent "soul," just as we must plainly be convinced that the single independent cell has a "psyche," we must as decidedly attribute a soul to every other cell; for the most important active constituent of the cell, the protoplasm, [58] everywhere exhibits the same psychic properties of sensibility or irritability, and motive power or will. The only difference is this, that in the organism of the higher animals and plants the numerous collected cells, to a great extent, give up their individual independence, and are subject, like good citizens, to the soul-polity which represents the unity of the will and sensations in the cell community. We here also must distinguish clearly between the central soul of the whole many-celled organism or the personal psyche (the person-soul), and the particular individual soul or elementary soul of the individual cells constituting that organism (the cell-soul). Their relations are strikingly illustrated in the instructive group of Siphonophora, as I have briefly shown in my article on "The Cell-soul and Soul-cells" (Deutsche Rundschau, July 1878). Beyond a doubt the whole stock or polity of Siphonophora has a very definite united will and a united sensibility, and yet each of the individual persons of which this stock (or Cormus) is composed has its own personal will and its own particular sensations. Each of these persons indeed was originally a separate Medusa, and the individual Siphonophora stock originated, by association and division of labour, out of these united Medusa communities.

When I developed this theory of the cell-soul and designated it in my Munich address as the "surest [59] foundation of empirical psychology," I believed I was drawing an inference quite to Virchow's mind, from his own views of mechanical and cellular-physiology; and for that reason I took the same occasion specially to celebrate his very great services to the cell theory. How astonished then was I when in his reply this very theory was violently attacked and satirised as "mere trifling with words." It never could have occurred to me that Virchow had long since become unfaithful to his most important biological principles, and had deserted his own mechanical "theory of cells;" it never had occurred to me that Virchow could be in great measure wanting in that zoological knowledge which is requisite for a practical comprehension of the cell-soul theory. He has never thoroughly studied either the one-celled Protozoa, the Infusoria and Lobosa, nor the Coelenterata, the highly instructive Sponges, Hydroids, Medusæ, or Siphonophora; and thus he is wanting in those genetic principles of comparative zoology on which our theory rests. It is in no other way conceivable that Virchow should contemn the most important consequences of the cell theory as "mere trifling with words."

Next to the one-celled infusoria no phenomenon throws such direct light on our cellular psychology as the fact that the human ovum, like the ova of all other animals, is a single, simple cell. In accordance [60] with our monistic conception of the cell-soul, we must conclude that the fertilised ovum-cell already virtually possesses those psychical properties which, by the special combination of the peculiarities inherited from both parents, characterise the individual soul of the new person; in the course of the development of the germ, the cell-soul of the fertilised ovum naturally is developed simultaneously with its material substratum, and subsequently, after birth, it appears in full activity.

According to Virchow's dualistic conception of the psyche, we must, on the contrary, assume that this immaterial essence at some period of its embryonic development (apparently when the spine separates itself from the external germ-layer) informs the soulless germ. Of course, the bare miracle is thus complete, and the natural and unbroken continuity of development is superfluous.

[61]

CHAPTER V.

THE GENETIC AND DOGMATIC METHODS OF TEACHING.

The very justifiable surprise which Virchow's Munich address has excited in many circles is due only in part to his opposition to the theory of descent; for the rest, and in much greater part, it is due to the astounding arguments which he has connected with it, particularly as to freedom for instruction. These arguments so closely resemble those of the Jesuits that they might have been inspired direct from the Vatican, or, which is the same thing, the notorious "court-chaplain party" in Berlin. No wonder, then, that these propositions, which would undermine the whole liberty of science, have met with the loudest approbation from the "Germania," the "New Evangelical Church Times" ("Neue Evangelischen Kirchenzeitung"), and other leading, equivocating organs of the Church militant. On the other hand, these odious principles are already so extensively discussed, and have been so clearly laid down in all their indefensibility, that I may here deal with them briefly. [62]

Virchow's politics as a pedagogue reach their highest pitch in this demand: "that in all schools, from the poor schools to the universities, nothing shall be taught that is not absolutely certain. None but objective and absolutely ascertained knowledge is to be imparted by the teacher to the learner; nothing subjective, no knowledge that is open to correction, only facts, no hypotheses." The investigation of such problems as the whole nation may be interested in must not be restricted; that is liberty of inquiry; but the problem ought not, without anything farther, to be the subject of *teaching*. "When we teach we must restrict ourselves to the smaller, and yet how great, departments which we are actually masters of."

Rarely indeed has such a treasonable attempt on liberty of doctrine been made by a prominent representative of science, and a leader of the intellectual movement too, as this by Virchow. Only inquiry is to be free and not teaching! And where in the whole history of science is there one single scientific inquirer to be found who would not have felt himself quite justified in teaching his own subjective convictions with as much right as he had to construct them from inquiry into objective facts. And where, generally speaking, is the limit to be found between objective [63] and subjective knowledge? Is there, in fact, any objective science?

This question Virchow answers in the affirmative, for he goes on to say: "We must not forget that there is a boundary line between the speculative departments of natural science and those that are actually conquered and firmly established" (p. 8). In my opinion, there is no such boundary line; on the contrary, all human knowledge as such is subjective. An objective science which consists merely of facts without any subjective theories is inconceivable. For evidence in favour of this view we must take a rapid survey of the whole domain of human science, and test the chief departments of it to see how far they contain, on the one hand, objective knowledge and facts, and on the other, subjective knowledge and hypotheses. We may begin directly with Kant's assertion that in every science only so much true—that is objective—knowledge is to be found as it contains of mathematics. Unquestionably mathematics stand at the head of all the sciences as regards the certainty of its teaching. But how as to those deepest and simplest fundamental axioms which constitute the firm basis on which the proud edifice of mathematical teaching rests? Are these certain and proved? Certainly not. The bases of its teaching are simply "axioms" which are incapable of proof. To give only one example of how the very [64] first principles of mathematics might be attacked by scepticism and shaken by philosophical speculation, we may remember the recent discussions as to the three dimensions of space and the possibility of a fourth dimension; disputes which are carried on even at the present day by the most eminent mathematicians, physicists, and philosophers. So much as this is certain, that mathematics as little constitute an absolutely objective science as any other, but by the very nature of man are subjectively conditioned. A man's subjective power of knowing can only discern the objective facts of the outer world in general so far as his organs of sense and his brain admit in his own individual degree of cultivation.

However, granting that mathematics practically constitute an absolutely certain and objective science, how is it with the rest of the sciences? Undoubtedly the most certain among them are those "exact sciences" whose principles are to be directly proved by mathematics; thus, in the first place, a great part of physics. We say, "a great part," for another large part—to speak accurately, by far the greatest—is incapable of any exact mathematical proof. For what do we know for certain of the essential nature of matter, or the essential nature of force? What do we know for certain of gravitation, of the attraction of mass, of its effects at great distances, and so on? [65] Newton's theory of gravitation is regarded as the most important and certain theory of physics, and yet gravitation itself is a hypothesis. Then, as to the other branches of physics—electricity and magnetism. The whole scheme of these important sciences rests on the hypothesis of "electric fluidity," or of imponderable matter of which the existence is nothing less than proved. Or optics? Optics certainly appertain to the most important and completest branch of physics, and yet the undulatory theory of light, which we accept now as the indispensable basis of optics, rests on an unproved hypothesis, on the subjective assumption of an ethereal medium, whose existence no one is in a position to prove objectively in any way. Nay, further, before Young set up the undulatory theory of light, for a hundred years the emanation theory as taught by Newton obtained exclusively in physics; a theory which at the present day is universally regarded as untenable. In our opinion the mighty Newton won the greatest honours in the development of the science of optics, inasmuch as he was the first to connect and explain the vast mass of objective optical facts by a subjective and pregnant hypothesis. But, according to Virchow's view, Newton on the contrary transgressed greatly by teaching this erroneous hypothesis; for even in "exact" physics none but "independent and certain facts" are to be taught and [66] established by "experiment as the highest means of proof."

Physics as a whole, as resting on mere unproved hypotheses, may be indeed an object of inquiry but not of teaching.

Of course the same is true of chemistry; nay, this stands on much weaker feet, and is even less proved than physics. The whole theoretical side of chemistry is an airy structure of hypotheses such as does not exist in any other science. In the last three decades we have seen a whole series of the most different theories rapidly succeed each other, none of which can be positively proved, though at least one of them is taught by every professor of chemistry. But what is worst of all, the common basis of all the most dissimilar chemical theories, viz., the atomic theory, is as unproved and unprovable as any hypothesis can be. No chemist has ever seen an atom, but he nevertheless considers the mechanism of atoms as the highest term of his science, he nevertheless describes and constructs the connection of atoms in their various combinations as though he had them before him on the dissecting-table! All the conceptions which we possess as to chemical structure and the affinities of matter, are subjective hypotheses, mere conceptions as to the position and changes of position of the various atoms, whose very existence is incapable of proof. Away, [67] then, with chemistry from our schools! The chemist must only describe the properties of the different elements and those combinations which can be put before the pupil as ascertained facts founded in experiment, "the highest means of proof." Everything that goes beyond this is mischievous, particularly every suggestion as to the essence and chemical constituents of bodies; matters as to which, in the nature of things, we can only form uncertain hypotheses. For as all chemistry, viewed as a system of doctrine, rests solely on such hypotheses, it may be indeed a subject of investigation but not of teaching.

Having thus convinced ourselves that chemistry as well as physics, those "exact sciences," those "mechanical" bases of all other sciences, rest on mere unproved hypotheses, and so must not be taught, we may make short work of the other faculties. For they collectively are more or less historical sciences and dispense wholly or in part with even those half-exact, fundamental principles on which physics and chemistry are based. In the first place, there is that grand, historical, natural science, geology; the great doctrine of the structure and composition, the origin and development of our globe. According to Virchow this too must be limited to the description of ascertained facts, such as the structure of mountain masses, the character of the fossils [68] they contain, the formation of crystals, and so forth. But not for the world must anything be taught as to the evolution of this globe; for this rests from beginning to end on unproved hypotheses. For even to the present day the Plutonic and Neptunic theories are disputing the field, and to this day we know not as to many of the most important rocks, whether they originated by the agency of fire or of water. The new and remarkable discoveries of the great Challenger-expedition threaten to subvert a great many geological notions which had long been regarded as certain. Then again, as to fossils. Who can prove with any certainty that these petrifactions are in truth the fossilised remains of extinct organisms? They may be—as many distinguished naturalists of even the last century maintained—marvellous sports of nature, mysterious "Lusus naturæ," or mere rough, inorganic models of the labouring Creator into which He subsequently "breathed the breath of life;" or perhaps "stone-flesh" (caro fossilis) brought into existence, on the dead rocks by the "fertilising air" (aura seminalis), and so forth.

But I am wrong! for with regard to petrifactions, Virchow is in the highest degree speculative, and accepts without any hesitation the rash hypothesis that fossils are actually the remains of extinct organisms, [69] although no "certain proof" whatever can be offered in its favour, and although experiment, the "highest means of proof," has never yet produced a single fossil. According to him these are actual "objective, material evidences," only here we must go no further than certain experience teaches us, and base no subjective conclusions on these objective facts. Thus, for instance, in the long series of the mesozoic formations, in the different strata of the Trias, Jurassic, and Chalk formations, for the deposition of which a lapse of many millions of years has been required, we find absolutely no remains of fossil mammalia beyond lower jaws; seek where we will, nothing is anywhere to be found but lower jaws, and no other bones whatever. The simple reasons of this striking imperfection of the palæontological record have been clearly expounded by Lyell, Huxley, and others. (Comp. my "History of Creation," vol. ii. p. 32.) These great investigators, in accordance with all other palæontologists, have demonstrated that these jaw-bones of the mesozoic period are the remains of mammalia, accurately speaking of marsupials, on the simple ground that the nether jaws of the extant recent marsupials show a similar characteristic form with the fossil ones. They therefore unhesitatingly assume that the rest of the bones in the bodies of these extinct animals corresponded to those of living mammals. [70] But this is a quite inadmissible hypothesis devoid of any "certain proof!" Where, then, are the other bones? Let us see them! till then we decline to believe in them. According to Virchow, we ought rather to assume that the lower jaw was the only bone in the body of these extraordinary beasts. Are there not, in fact, snails, in which an upper jaw is the only representation of a skeleton.

We cannot omit taking this opportunity of casting a side glance at the very hazardous position which Virchow, in total opposition to his boasted cool scepticism, has taken up in anthropology as it is called, now his favourite branch of science. In his Munich address he tells us that he is pursuing the study of anthropology with delight, and then asserts that "the quarternary man" is an universally-accepted fact. Quite apart from this statement, we have seen that Virchow can never attain to a profound and really scientific study of anthropology simply for this reason, that he is lacking in that comprehensive knowledge of comparative morphology which is indispensable to it; nay, comparative anatomy and ontogenesis must be, according to him, unpermitted speculations and the phylogenesis of man, the key to all the most important questions of anthropology, being based upon these, is devoid of all certain proof. All the more must we wonder at the [71] speculative levity with which even the sceptic Virchow in the "Primeval History of Man" and "Fossil Anthropology," embarks in the most hazardous conjectures, and gives out uncertain, subjective hypotheses as certain, objective facts.

There is, in fact, at the present day no department of science in which the wildest and most untenable hypotheses have blossomed out so freely as in anthropology and ethnology, so-called. All the phylogenetic hypotheses which I myself have put forward in my "Evolution of Man" as to the animal ancestry of man, or in my "Natural History of Creation" as to the affinities of animal races — all the other genealogical hypotheses which are now advanced by numerous zoologists and botanists as to the phylogenetic evolution of the animal and plant worlds—all these hypotheses together, which Virchow rejects in a lump, are, critically considered as hypotheses, far better grounded in facts, far better supported by facts, than the majority of those innumerable airy and fanciful hypotheses with which, for the last twelve years, the "Archiv für Anthropologie" and "Zeitschrift für Ethnologie," edited by Virchow and Bastian, have filled their columns. This last periodical has at least the merit of being a tolerably consistent opponent of the doctrine of evolution, while in the former, during twelve years, essays on both sides have been mixed up in cheerful confusion. [72] And how fanciful are the short-sighted hypotheses which there blossom forth from the mixed mass of facts, chaotically flung together. Only think of the disputes over the stone age, bronze age, and iron age; think of the motley discussions as to the varieties of skull-conformation and their significance; on the races of man, the migrations of peoples and the like. Most of these very intricate historical problems are far more buried in obscurity, and the hypotheses to explain them dispense far more largely with any basis of facts, than is the case with our phylogenetic hypotheses; for these are more or less "objectively" based on the facts of comparative anatomy and ontogenesis.

But no one of these historical hypotheses is so daring, so little "certainly proved," as the group of very various and contradictory hypotheses which have been put forward as to the antiquity and first appearance of the human species; and Virchow asserts positively "The pleistocene man is an universally accepted fact. The tertiary man is, on the other hand, a problem, though indeed a problem which is already under substantial discussion!" As if the distinction between the tertiary and quaternary periods were not itself a geological hypothesis, and as if the significance of the fossil animal-remains, which play the largest part in it, did not also rest on mere hypotheses which escape all certain proof! Where, then, is the [73] actual experiment "as the highest means of proof," which gives evidence for these "certain facts"? The whole discussion in general about pre-historic man, which Virchow has mixed up with his Munich address (pp. 30, 31), is the clearest evidence of the uncritical spirit in which he deals with these historical problems as "exact natural sciences." He assures us that "not one single ape's skull, nor skull of an anthropoid ape, has ever been found which could actually have belonged to a human owner! and he adds this sentence, in italics, "We cannot teach, for we cannot regard it as a real acquisition of science, that man is descended from the ape or from any other animal!" Then evidently no alternative remains but that he is descended from a god, or from a clod!

But let us go over the rest of the sciences to see what, according to Virchow, may be taught in each without endangering the safety of science. In the whole department of biology, as well as in zoology — including anthropology — and in botany, instruction must be limited to imparting those trifling fragments of knowledge which either consist of mere descriptions of dry facts, or which supply an explanation of them by mathematical formulas. Morphology must be taught as mere descriptive anatomy and systematising, the history of development as mere descriptive ontogenesis. Comparative anatomy and phylogenesis, which by their [74] explanatory hypotheses raise those dead masses of facts to the place of true and living sciences — these must not be taught at all. And how then do matters stand with regard to the cell-theory, that fundamental theory on which every element of our morphology and physiology depends, and by applying which Virchow himself reached his grandest results?

Since Schleiden in Jena, forty years ago, first put forward the cell-theory, and Schwann immediately after applied it to the animal kingdom and so to the whole organic world, this fundamental doctrine has undergone very important modifications, for it is indeed a biological theory, but not a fact. We may recollect under what different aspects its main principles have appeared in the course of these four decades: what changes have taken place in the conception of the cell itself. After the organic cell had originally been conceived of as a vesicle, consisting of a firm capsule and a fluid content, we subsequently discerned it to be composed of a glutinous semi-fluid cell-substance, the protoplasm, and convinced ourselves that this protoplasm and the cell-core or nucleus enclosed in it are the most important and indispensable constituent parts of the cell, while the external firm capsule, the cell-membrane, is not essential and very frequently wanting. But even now opinions widely differ as to how the conception of a cell should be precisely defined, [75] and what consequences must be inferred from the cell-theory, and attempts have not been wanting to upset it altogether and to treat it as worthless. The anatomist Henle, of Göttingen, in particular, has repeatedly made such an attempt, that "gifted" anatomist who, in the preface to his bulky text-book of human anatomy, declared that scientific ideas are mere worthless paper money, and that the noble metal of facts, on the contrary, is the only genuine article. Not long since a bulky volume in quarto appeared, by one Herr Nathusius-Königsborn, in which the cell is explained to be a subordinate plastic element, and the cell-theory is eliminated as superfluous; and this monstrous volume, full of the most amusing nonsense, is dedicated to Herr Henle. Virchow formerly was one of the victorious opponents of the Göttingen physician, and wrote brilliant articles against the "rational pathology" of "irrational Herr Henle;" now apparently he agrees with him that the paper money of ideas is worthless as compared with the noble metal of facts. Of course the cell-theory then loses all its value, and cannot be a subject of instruction; for the cell itself is not a certain and undoubted fact, but only an abstraction, a philosophical idea.

Nothing more clearly shows what a complete change Virchow has undergone in his most important principles, and what an utter metapsychosis in this special province, [76] than his famous axiom, uttered in 1855 — "Omnis cellula e cellula." That is unquestionably the boldest generalisation to which the youthful, independent Virchow ever attained, and one on which he justly prided himself not a little. He himself repeatedly compared it with Harvey's saying, which marked an epoch — "Omne vivum ex ovo." But neither of these axioms is universally correct. On the contrary, we now know that every cell does not necessarily originate from a cell, any more than that every organic individual originates from an ovum. In many cases true nucleated cells proceed from un-nucleated cytods, as in the Gregarinæ, Myxomycetæ and others. Nay more, the primordial organic cells could only have originated in the first instance from non-cellular plastides or monads by their homogeneous plasson resolving itself into an internal nucleus and an external protoplasm. Thus, as we subsequently learnt to know most of the exceptions to this generalisation of Virchow, it appeared all the bolder; the more so as we were at that time far from being able to refer all the different tissues of the higher animals with any certainty to cells, and as not a few experiments seemed to point to the hypothesis of free cell-formation. That guiding axiom, which so powerfully furthered the cell-theory, Virchow, from his present standpoint, must wholly condemn as a crime against [77] exact science, and he surely can never forgive himself for having propounded this hypothesis—which was afterwards found to be not universally true—as an important doctrinal axiom.

We shall indeed find much worse sins against his own principles of to-day if we turn to Virchow's own special department of science, namely, pathological anatomy and physiology, the most important division of theoretic medicine. The great and incomparable services which Virchow here effected do not depend on the numerous independent new facts which he discovered, but on the theories and hypotheses by which, like an inspired pioneer, he sought to open a way through the dead waste of pathological knowledge and to form it into a living science. These new theories and the hypotheses on which they were founded, Virchow then propounded to us, his disciples, with such incisive assurance that every one of us was convinced of their truth; and yet later experience has shown that they were in part insufficiently proved and in part wholly false. For example, I will only here recall his famous theory of the connective-tissue, for which I myself in several of my early works (1856 to 1858) broke a lance. His theory seemed to explain a host of the most important physiological and pathological phenomena in the simplest manner, and yet it was afterwards proved to be false. In spite of [78] this, I declare to this day that it was of the greatest service for the development of our acquaintance with the formation of the connective-tissue; as a guiding hypothesis and as a provisional clue to our investigations. Virchow, on the contrary, if he impartially reflects on the part he took in the diffusion of this misleading doctrine, must reproach himself severely for it. For "we must draw a hard and fast line between what we are to teach and what we are to investigate. What we investigate are problems," but "the problem ought not to be the subject of teaching." That Virchow, in his course of instruction, every day belied this, his present view of teaching, that he every hour taught his disciples some unproved theory and problematical hypothesis, every one knows who, like myself, for years and with the deepest interest, enjoyed his distinguished instruction.

Still the captivating charm of this instruction—in spite of the defective method of unprepared lectures—lay precisely in this, that Virchow as a teacher constantly let us, his pupils, enter into those problems with which he himself at the moment was occupied; that he propounded to us his personal hypothesis for the elucidation of the given facts. And what really gifted teacher who lives in his science would not do the same? Where is there, or where has there ever been, a great master who in his teaching has confined himself to [79] only imparting certain and undoubtedly ascertained facts? Who has not, on the contrary, found that the charm and value of his instruction lay precisely in propounding the problems which link themselves with those facts, and in teaching the uncertain theories and fluctuating hypotheses which may serve to solve these problems? Or is there for the young and struggling mind anything better, or more conducive to culture, than to exercise the intelligence in problems of investigation?

How unpractical and how absurd is Virchow's demand—that only ascertained facts and no problematic theories shall be admitted in teaching—will be still more strikingly shown by a glance over the remaining provinces of human knowledge. What, indeed, will be left of history, of philology, of political science, of jurisprudence, if we restrict the teaching of them to absolutely-ascertained and established facts. What of "science" will remain to them if the idea which endeavours to discern the causes of the facts is banished? if the problems, the theories, the hypotheses, which seek these causes may not be generally taught? And that philosophy—the science of knowing—by which all the common results of human knowledge are to be bound up into one grand and harmonious whole—that philosophy, I say, must not be generally taught, is, according to Virchow, quite self-evident. [80]

Finally, there remains nothing but theology. Theology alone is the one true science, and its dogmas alone may be taught as certain. Of course! for it proceeds directly from revelation, and only divine revelation can be "quite certain;" it alone can never err. Yes, incredible as it sounds, Virchow, the sceptical opponent of dogma, the leader of the fight for "liberty of science," Virchow now finds the only sure basis for instruction in the dogmas of the Church. After all that has gone before, the following memorable sentence leaves no doubt on this score: — "Every attempt to transform our problems into dogmas, to introduce our conjectures as a basis of instruction, particularly any attempt simply to dispossess the Church and to supplant her dogma by a creed of descent—ay, gentlemen—this attempt must fail, and in its ruin will entail the greatest peril on the position of science in general."

The shouts of triumph of the whole clerical press over Virchow's Munich address is thus rendered perfectly intelligible, for it is well known that "there is more joy in heaven over one sinner that repenteth than over ten just men." When Rudolf Virchow, the "notorious materialist," the "advanced radical," the "great supporter of the atheism of science," is so suddenly converted, when he proclaims loudly and publicly that the dogmas of the Church are the only sure basis of instruction, then the Church militant may well sing [81] "Hosanna in the highest!" Only one thing is to be regretted, that Virchow has not more clearly defined which of the many different church-religions is the only true one, and which of the innumerable and contradictory dogmas are to form the sure basis of instruction. We all know that each Church regards itself as the only truly saving one, and her own dogma as the only true one. But as to whether it is to be Protestantism or Catholicism, the Reformed or the Lutheran confession, whether the Anglican or the Presbyterian dogma, whether the Roman or the Greek Church, the Mosaic or the Mohammedan dispensation, whether Buddhism or Brahmanism, whether, finally, it is to be one of the many fetish-religions of the Indians and Negroes that is to form the permanent and sure basis of instruction, let us hope that Virchow will at the next meeting of German naturalists and physicians divulge his opinion.

At any rate, the "instruction of the future, according to Virchow," will be greatly simplified if he will do this. For the dogma of the Trinity in Unity as a basis of mathematics, the dogma of the resurrection of the body as a basis of medicine, the dogma of infallibility as a basis of psychology, the dogma of the immaculate conception as a basis of genetic science, the dogma of the staying of the sun as a basis of astronomy, the dogma of the creation of the earth, animals, and plants [82] as a basis of geology and phylogenesis—these or any other dogma, at pleasure, from any other church will make all other doctrine quite superfluous. Virchow, "that critical spirit," knows as well as I, and as every other naturalist, that these dogmas are not true, and nevertheless, in his opinion, they are not to be supplanted as the "basis of instruction" by those theories and hypotheses of modern natural science of which Virchow himself says that they may be true, that in a great measure they probably are true, but are not yet "quite certainly proved."

At pages 15, 24, 26, 28, and elsewhere in his Munich address, Virchow strongly insists that only that objective knowledge may be taught which we possess as absolutely certain fact! and then at page 29 he requires us to conclude that the basis of instruction shall continue to be the purely subjective dogmas of the Church; revelations and dogmas which not only are not proved by any facts whatever, but on the contrary, stand in the most trenchant contradiction to the most obvious facts of natural experience and fly in the face of all human reason. These contradictions, to be sure, are no greater than some others which stand out conspicuous and incomprehensible in Virchow's discourse. Thus at the beginning of his address he glorifies Lorenz Oken and deeply laments "that he, that highly-valued and honoured master, that ornament [83] of the high school of Munich, had been forced to die in exile! That cruel exile which oppressed Oken's latter years, which left him to perish far from those cities to which he had sacrificed the best powers of his life, that exile will be remembered as the note of the time which we have passed through. And so long as there continue to be meetings of German naturalists, so long may we gratefully remember that this man to his death bore upon him all the signs of a martyr, so long shall we point to him as one of the witnesses who have fought for us and for the liberty of science." Verily these words from Virchow's lips sound like the bitterest irony; for was not Lorenz Oken one of the foremost and most zealous champions of that monistic doctrine of development against which Rudolf Virchow at this day is most violently striving? Did not Oken himself proceed farther in the construction of bold hypotheses and comprehensive theories than any supporter of the doctrine of evolution at the present time? Is not Oken justly considered as the one typical representative of that older period of natural philosophy who rose to much higher and bolder flights of fancy, and left the solid ground of facts much farther behind him than any tyro of the new philosophy? And this makes the irony seem all the greater with which Virchow at the beginning of his address [84] glorifies Oken the free teacher, as a martyr to the freedom of science, and at the end of it insists that this freedom applies only to inquiry and not to teaching, and that the master must teach no problem, no theory, no hypothesis.

While this unheard-of demand sets Virchow's views of teaching in the most extraordinary light, and while every unprejudiced and experienced teacher must most emphatically protest against this strait-waistcoat for instruction, he will feel no less bound to resist Virchow's other strange demand, that every ascertained truth shall forthwith be taught in all schools, down to the elementary schools. I myself, in my Munich address, sought the instructional value of our monistic evolution theory above all in the genetic method, in the inquiry, that is to say, for the effective causes of the facts taught; and I added these words—"How far the principles of the doctrine of universal evolution ought to be at once introduced into our schools, and in what succession its most important branches ought to be taught in the different classes—cosmogony, geology, the phylogenesis of animals and plants, and anthropology—this we must leave to practical teachers to settle. But we believe that an extensive reform of instruction in this direction is inevitable, and will be crowned by the fairest results." I purposely avoided [85] any closer discussion of this specialist question, as I felt not even approximately capable of solving it, and I believe, in fact, that none but skilled and experienced practical teachers can undertake the solution of it with any success.

For Virchow these specialist difficulties seem not to exist; he regards my reticence as a mere "postponement of the task," and he answers in the following astonishing sentences:—"If the theory of descent is as certain as Herr Haeckel assumes, then we must demand—for it is a necessary consequence—that it shall be taught in schools. How is it conceivable that a doctrine of such importance, which must effect such a total revolution in all our mental consciousness, which directly tends to create a new kind of religion, should not be included in the school scheme of instruction? How is it possible that such a—revelation, shall I say—should be in any measure suppressed, or that the promulgation of the greatest and most important advance which has been made in our views during the present century should be left to the discretion of schoolmasters? Ay, gentlemen, that would indeed be a renunciation of the hardest kind, and practically it could never be carried out! Every schoolmaster who assumes this doctrine for himself will involuntarily teach it, how can it be otherwise?" [86]

I must here be permitted to take Virchow exactly at his word. I endorse almost all that he has said in these and the following sentences. The only difference in our views is this, that Virchow regards the theory of descent as an unproved and unproveable hypothesis; I, on the contrary, as a fully established and indispensable theory. How then will it be if the teachers of whom Virchow speaks agree with my views, if—apart, of course, from all special theories of descent—they, like me, consider the general theory of descent as the indispensable basis of all biological teaching? And that that is actually the case Virchow may easily convince himself if he looks over the recent literature of zoology and botany! Our whole morphological literature in particular is already so deeply and completely penetrated by the doctrine of descent, phylogenetic principles already prevail so universally as a certain and indispensable instrument of inquiry, that no man for the future would deprive himself of their help. As Oscar Schmidt justly observes—"Perhaps ninety-nine per cent. of all living, or rather of all working zoologists, are convinced by inductive methods of the truth of the doctrine of descent." And Virchow with his magisterial requirements will attain only the very reverse of what he aims at. How often has it not been said already that science must either have perfect freedom or else none at all? This is as true [87] of teaching as it is of inquiry, for the two are intrinsically and inseparably connected. And so it is not in vain that it is written in section 152 of the German Code, and in section 20 of the Prussian Charter, "Science and her teaching shall be free!"

[88]

CHAPTER VI.

THE DOCTRINE OF DESCENT AND SOCIAL DEMOCRACY.

Every great and comprehensive theory which affects the foundations of human science, and which, consequently, influences the systems of philosophy, will, in the first place, not only further our theoretical views of the universe, but will also react on practical philosophy, ethics, and the correlated provinces of religion and politics. In my paper read at Munich I only briefly pointed out the happy results which, in my opinion, the modern doctrine of evolution will entail when the true, natural religion, founded on reason, takes the place of the dogmatic religion of the Church, and its leading principle derives the human sense of duty from the social instincts of animals.

The references to the social instincts which I, in common with Darwin and many others, regard as the proper source and origin of all moral development, appear to have afforded Virchow an opportunity in his reply for designating the doctrine of inheritance as a "socialist theory," and for attributing to it the most [89] dangerous and objectionable character which, at the present time, any political theory can have; and these startling denunciations so soon as they were known called forth such just indignation and such comprehensive refutation that I might very properly pass them over here. Still we must at least shortly examine them, in so far as they supply a further proof that Virchow is unacquainted with the most important principles of the development-theory of the day, and therefore is incompetent to judge it. Moreover, Virchow, as a politician, manifestly attributed special importance to this political application of his paper, for he gave it the title, which otherwise would have been hardly suitable, of "The Freedom of Science in the Modern Polity." Unfortunately he forgot to add to this title the two words in which the special tendency of his discourse culminates; the two pregnant words, "must cease!"

The surprising disclosures in which Virchow denounces the doctrine of evolution, and particularly the doctrine of descent, as socialist theories and dangerous to the community, run as follows:—
"Now, picture to yourself the theory of descent as it already exists in the brain of a socialist. Ay, gentlemen, it may seem laughable to many, but it is in truth very serious, and I only hope that the theory of descent may not entail on us all the horrors which similar theories have [90] actually brought upon neighbouring countries. At all

times this theory, if it is logically carried out to the end, has an uncommonly suspicious aspect, and the fact that it has gained the sympathy of socialism has not, it is to be hoped, escaped your notice. We must make that quite clear to ourselves."

On reading this statement, which seems extracted from the Berlin "Kreuz-Zeitung," or the Vienna "Vaterland," I ask myself in surprise, "What in the world has the doctrine of descent to do with socialism?" It has already been abundantly proved on many sides, and long since, that these two theories are about as compatible as fire and water. Oscar Schmidt might with justice retort, "If the socialists would think clearly they would feel that they must do all they can to choke the doctrine of descent, for it declares with express distinctness that socialist ideas are impracticable." And he proceeds to add, "And why has not Virchow made the gentle doctrines of Christianity responsible for the excesses of socialism? That would have had some sense. His denunciation flung so mysteriously and so confidently before the great public, as though it concerned 'a sure and attested scientific truth,' is, at the same time, so hollow that it cannot be brought into harmony with the dignity of science."

With all these empty accusations, as with all the empty reproaches and groundless objections which [91] Virchow brings against the doctrine of evolution, he takes good care in no way to touch the kernel of the matter. How, indeed, would it have been possible without arriving at conclusions wholly opposed to those which he has declared? For the theory of descent proclaims more clearly than any other scientific theory, that that equality of individuals which socialism strives after is an impossibility, that it stands, in fact, in irreconcilable contradiction to the inevitable inequality of individuals which actually and everywhere subsists. Socialism demands equal rights, equal duties, equal possessions, equal enjoyments for every citizen alike; the theory of descent proves, in exact opposition to this, that the realisation of this demand is a pure impossibility, and that in the constitutionally organised communities of men, as of the lower animals, neither rights nor duties, neither possessions nor enjoyments have ever been equal for all the members alike nor ever can be. Throughout the evolutionist theory, as in its biological branch, the theory of descent—the great law of specialisation or differentiation—teaches us that a multiplicity of phenomena is developed from original unity, heterogeneity from original similarity, and the composite organism from original simplicity. The conditions of existence are dissimilar for each individual from the beginning of its existence; even the inherited qualities, the natural "disposition," are [92] more or less unlike; how, then, can the problems of life and their solution be alike for all? The more highly political life is organised, the more prominent is the great principle of the division of labour, and the more requisite it becomes for the lasting security of the whole state that its members should be variously distributed in the manifold tasks of life; and as the work to be performed by different individuals is of the most various kind, as well as the corresponding outlay of strength, skill, property, &c., the reward of the work must naturally be also extremely various. These are such simple and tangible facts that one would suppose that every reasonable and unprejudiced politician would recommend the theory of descent, and the evolution hypothesis in general, as the best antidote to the fathomless absurdity of extravagant socialist levelling.

Besides, Darwinism, the theory of natural selection—which Virchow aimed at in his denunciation, much more especially than at transformation, the theory of descent—which is often confounded with it—Darwinism, I say, is anything rather than socialist! If this English hypothesis is to be compared to any definite political tendency—as is, no doubt, possible—that tendency can only be aristocratic, certainly not democratic, and least of all socialist. The theory of selection teaches that in human life, as in animal and plant [93] life everywhere, and at all times, only a small and chosen minority can exist and flourish, while the enormous majority starve and perish miserably and more or less prematurely. The germs of every species of animal and plant and the young individuals which spring from them are innumerable, while the number of those fortunate individuals which develop to maturity and actually reach their hardly-won life's goal is out of all proportion trifling. The cruel and merciless struggle for existence which rages throughout all living nature, and in the course of nature *must* rage, this unceasing and inexorable competition of all living creatures, is an incontestable fact; only the picked minority of the qualified "fittest" is in a position to resist it successfully, while the great majority of the competitors must necessarily perish miserably. We may profoundly lament this tragical state of things, but we can neither controvert it nor alter it. "Many are called but few are chosen." The selection, the picking out of these "chosen ones," is inevitably connected with the arrest and destruction of the remaining majority. Another English naturalist, therefore, designates the kernel of Darwinism very frankly as the "survival of the fittest," as the "victory of the best." At any rate, this principle of selection is nothing less than democratic, on the contrary, it is aristocratic in the strictest sense of the word. If, therefore, [94] Darwinism, logically carried out, has, according to Virchow, "an uncommonly suspicious aspect," this can only be found in the idea that it offers a helping hand to the efforts of the aristocrats. But how the socialism of the day can find any encouragement in these efforts, and how the horrors of the Paris Commune can be traced to them, is to me, I must frankly confess, absolutely incomprehensible.

Moreover, we must not omit this opportunity of pointing out how dangerous such a direct and unqualified transfer of the theories of natural science to the domain of practical politics must be. The highly elaborate conditions of our modern civilised life require from the practical politician such circumspect and impartial consideration, such thorough historical training and powers of critical comparison, that he will not venture to make such an application of a "natural law" to the practice of civilised life, but with the greatest caution and reserve. How, then, is it possible that Virchow, the experienced and skilled politician, who, above all things, preaches caution and reserve in theory, suddenly makes just such an application of transformation and Darwinism — an application so radically perverse that it actually flies in the face of the fundamental ideas of these doctrines? I myself am nothing less than a politician. In direct contrast with Virchow, I lack alike the gift and the training for [95] it, as well as taste and vocation. Hence I neither shall play any political part in the future, nor have I hitherto made any attempt of the kind. Though here and there I have occasionally uttered a political opinion, or have made a political application of some theory of natural science, these subjective opinions have no objective value. In point of fact I have by so doing overstepped the limits of my competence, just as Virchow has by going into questions of zoology and particularly that of the transformation of apes: I am a layman in political practice, as Virchow is in the province of zoological hypothesis. Moreover, such success as Virchow has attained during the twenty years of his painful, wearisome, and exhausting activity as a politician does not, in truth, make me pine for such laurels.

But this at least I, as a theoretical naturalist, may demand of practical politicians, that in utilising our theories for political ends they should first make themselves exactly acquainted with them; they then, for the future, would forbear drawing conclusions from them, the very opposite to those which ought reasonably to be inferred. Misunderstandings would never thus be wholly avoided, it is true, but what doctrine is universally secure against misunderstanding? And from what theory, however sound and true, may not the most unsound and frantic inferences be drawn? [96]

Nothing, perhaps, shows so plainly as the history of Christianity how little theory and practice harmonise in human life; how little pains are taken, even by those whose calling it is to uphold established doctrines, to apply their natural consequences to practical life. The Christian religion, no doubt, as well as the Buddhist, when stripped of all dogmatic and fabulous nonsense, contains an admirable human kernel, and precisely that human portion of Christian teaching—in the best sense social-democratic—which preaches the equality of all men before God, the loving of your neighbour as yourself, love in general in the noblest sense, a fellow-feeling with the poor and wretched, and so forth—precisely, those truly human sides of the Christian doctrine are so natural, so noble, so pure, that we unhesitatingly adopt them into the moral doctrine of our monistic natural religion. Nay, the social instincts of the higher animals on which we found this religion (for instance the marvellous sense of duty of ants, &c.) are in this best sense strictly Christian.

And what—we may ask—what have the professed supporters, the "learned divines" of this religion of love done? Their deeds are written in letters of blood in the history of the civilisation of mankind during the last 1800 years. All else that differing church-religions have accomplished for the forcible extension of their doctrines and for the extirpation of heretics of [97] other creeds, all that the Jews have been guilty of towards the heathen, the Roman emperors towards the Christians, the Mohammedans towards Christians and Jews alike—all this is outdone by the hecatombs of human victims which Christianity has demanded for the spread of her doctrines. And these were Christians against Christians—orthodox Christians against heterodox Christians! think only of the Inquisition in the Middle Ages, of the inconceivable and inhuman barbarities committed by the "most Christian kings" of Spain, by their worthy colleagues in Frankfort, in Italy, and elsewhere. Hundreds of thousands then died that most horrible death by fire, simply because they would not bend their reason to pass under the yoke of the grossest superstition, and because their loyalty to their convictions forbade them to deny the natural truth that they clearly discerned. There are no deeds more hideous, base, and inhuman than those that at that time were committed—nay, are still committed—in the name and on account of "true Christianity."

And finally, how do matters stand with regard to the morality of the priests who announce themselves as the ministers of God's Word, and whose duty is therefore above all others to carry out the saving doctrines of Christianity in their own lives? The long, unbroken, and horrible series of crimes of every kind which is offered by the history of the Roman Popes is [98] the best answer to this question. And just as these "Vicars of God on earth" did, so did their subordinates and accomplices, so, too, have the orthodox priests of other sects done; never failing to set the practice of their own course of life in the strongest possible contrast to those noble doctrines of Christian love which were constantly on their lips.

And as with Christianity so it is with every other religious and moral doctrine which ought to have proved its power in the wide domain of practical philosophy, in the education of youth, in the civilisation of nations. The theoretic kernel of this doctrine may always and everywhere stand in the most glaring contradiction to its practical working-out, testifying to the endless inconsistency of human nature: but what can all this matter to the scientific inquirer? His sole and only task is to seek for truth and to teach what he has discerned to be the truth, indifferent as to what consequences the various parties of state or church may happen to draw from it.

[99]

CHAPTER VII.

IGNORABIMUS ET RESTRINGAMUR.

The dangerous attempt which Virchow made in Munich against the freedom of science is not the first of its kind. On the contrary, five years before, it experienced a similar attack which is most intimately connected with this later one, so that, in conclusion, we must here add a few words on the subject. Undoubtedly the famous "Ignorabimus-speech" of Du Bois-Reymond, which he delivered in 1872 at the forty-fifth meeting of German naturalists and physicians in Leipzig, forms only the first portion of that same crusade against the freedom of science of which Virchow's "Restringamur speech" of 1877, at the fiftieth meeting of the same society, forms the second part.

That brilliant and powerful essay by Du Bois-Reymond "on the Limitation of Natural Knowledge" has already been discussed so often, and from such different sides, that it might seem superfluous to say another word about it. It seems to me, nevertheless, that by most people the centre-of-gravity of its contents was [100] overlooked in admiration of the brilliant accessories of the essay. Indeed this frequently happens with Du Bois-Reymond's articles, for he knows too well how to conceal the weakness of his argument and evidence, and the shallowness of his thought, by striking images and flowery metaphors, and by all the phraseology of rhetoric in which the versatile French nature is so superior to our sober German one. It is all the more important that we should not let ourselves be dazzled by these seductive tricks, and particularly by adduced facts which bear upon the most important and fundamental questions of human science, but that we should extract the hard kernel from the savoury and fragrant fruit. In the preface to my "Evolution of Man," and in the notes 22 and 23 of my Munich address, I have already incidentally alluded to the chief weaknesses of the "Ignorabimus-speech;" but I must here return somewhat more fully to the subject.

There are, as is well known, two problems which Du Bois-Reymond propounds as the impassable boundary of human knowledge of nature; limits which indeed the human mind is not only incapable of passing at the present stage of its development, but which it never can be capable of passing in any more advanced stage. The first problem is the nature and connection of matter and force; the second is human consciousness. Now, first of all, as has already been said in [101] the preface to the "Evolution of Man," we must raise a decided protest against the air of infallibility with which Du Bois-Reymond pronounces that these two problems are insoluble, not only at the present time but to all futurity. The power of development inherent in science and knowledge is hereby simply swept away with a word. Almost every great and difficult problem of knowledge seems to most or all contemporary thinkers insoluble, and every path to the solution of it seems closed, till at last the bold genius appears whose clear sight detects the right path which till then was hidden, and which leads to the required knowledge. We need only call to mind our present doctrine of evolution. The problem of creation — the question as to the origin of animal and vegetable species — was universally looked upon as transcendental and perfectly insoluble, till the genius of Lamarck established the principles of the theory of descent in his admirable "Philosophie Zoologique" in 1809. Nay, even then most — and among them the most distinguished — biologists thought the problem of creation a quite insoluble mystery, and Darwin was the first to solve it, fifty years later, by his theory of selection in 1859. Hence we venture to assert that there is no scientific problem of which we may dare to say that the mind of man will never solve it even in the remotest future. Well does Darwin [102] say, in the introduction to his "Descent of Man," "Ignorance more frequently begets confidence than does knowledge: it is those who know little and not those who know much who so positively assert that this or that problem will never be solved by science." As far as concerns the two separate limits which Du Bois-Reymond fixes for human knowledge, in my opinion they are undoubtedly identical. The problem of the origin and nature of consciousness is only a special case of the general problem of the connection of matter and force. Du Bois-Reymond himself indicates that this is possible at the close of his paper; for he says, "Finally, the question arises whether the two limitations to our natural knowledge may not perhaps be identical; that is to say, whether if we could conceive of the true essence of matter and force, we should not also understand how the substance which lies at their root can, under certain given conditions, feel, desire, and think. This conception is, no doubt, the simplest, and according to admitted principles of inquiry it is to be preferred to that other which it confutes, and according to which, as has been said, the world appears doubly incomprehensible. But it is in the very nature of things that we cannot on this point come to any clear conclusion, and all further words on the subject are idle — and so, "Ignorabimus."

The light way in which Du Bois-Reymond here [103] passes over the most important part of his subject is truly surprising; as if it were ultimately indifferent whether we have before us one single insoluble fundamental problem or two quite different ones; and as if mature reflection did not lead to the conviction that, in fact, the second problem is only a special case of the first general problem. I, for my part, cannot conceive of them in any other relation; I think, too, that all further words are by no means superfluous, but on the contrary conduce to a very strong conviction of the unity of the problem. That Du Bois-Reymond also has not come to any clear conclusion on this point lies, not alone in the "nature of things," but, as in Virchow's case, in the nature of the investigator himself; in his lack of knowledge of the history of evolution, and in his neglect of those comparative and genetic methods of study, without which, in my opinion, not even an approximate solution of this highest and most difficult question is to be looked for.

Nothing appears to me to be of more importance for the mechanical explanation of consciousness than the comparative consideration of its development. We know that a new-born child has no consciousness, but that it is slowly and gradually acquired and developed. We perceive for ourselves how unconscious actions become conscious, and *vice versa*. Innumerable actions [104] which at first are troublesome and have to be learnt with consciousness and reflection — as for instance walking, swimming, singing, and so forth — become unconscious only by repetition, practice, and the habit of using the organs. On the contrary, unconscious actions become conscious as soon as we direct our attention to them or our self-observation is attracted to them; as for instance when we miss a step in going up stairs or touch a wrong note on the piano; and beyond a doubt, conscious and unconscious actions pass into each other without any distinct line of demarcation. Finally, we see no less plainly by a comparative consideration of the soul-life of animals, that their consciousness is slowly, gradually, and serially developed, and that a long unbroken series of steps leads from unconscious to conscious existence. From these comparative and genetic experiences we may draw the conclusion that consciousness, like sensation and volition, like all the other soul-activities, is a function of the organism, a mechanical activity of the cells; and, as such, is referable to chemical and physical processes. Hence, if we were in a position to understand force as a necessary function of matter, we could explain consciousness, as well as the soul in general, as a necessary function of certain cells.

How little Du Bois-Reymond is acquainted with the facts of comparative and genetic psychology, nothing [105] shows more strikingly than the following astounding proposition in the "Ignorabimus-speech:"—"Where the material conditions for psychical activity, in the form of a nervous system, are wanting, as in plants, the naturalist cannot recognise a soul-life, and, on this point, he but seldom meets with contradiction." Begging your pardon! Every naturalist who is familiar with the comparative morphology and physiology of the lower animals will here put in a decided contradiction, for he can no more refuse to admit the undoubted sensation and voluntary motion of the one-celled Infusoria than of the many-celled hydroid polyps. The body of the true Infusoria (Ciliata, Acineta, &c.), and many other Protista, remain throughout life one single cell, and, nevertheless, this cell is as fully furnished with all the most important attributes of the soul, with sensation and volition, as any one of the higher animals with a nervous system. The same obtains of the Hydra and the related hydroid polyps, in which the neuro-muscular cells, or other distributed cells of the outer germ-layer, fulfil the soul-functions. But as these cells, besides this, exercise motor and other functions as well, we cannot as yet designate them as nerve-cells, at any rate there can be no idea of a special nervous-system. The characteristic soul-organs of the higher animals, which we include under the conception of a nervous-system, in fact originated [106] by the division of labour of the cells out of those neutral cell-groups in their lower-typed ancestors.

In the great Soul-question Du Bois-Reymond, like Virchow, still keeps his position on the standpoint of neural-psychology, according to which no personal soul-life is conceivable without a nervous system. We look upon this standpoint as left far behind, and set up in opposition to it Cellular-psychology, the doctrine that every animal cell has a soul; that is to say, that its protoplasm is endowed with sensation and motion. In the one-celled Infusoria, which are so highly sensitive and have such an energetic will, this conception will be clear without any farther explanation. But we cannot refuse to allow that plant-cells as well as animal-cells have psychic functions, since we know that the phenomena of irritability, and of "automatic motion," are the universal attributes of all protoplasm. No doubt the specific mechanism, the cause of motion, in the irritable Mimosa and other "sensitive" plants, is quite different from the muscular motions of animals; but these, like those, are only specifically different forms of development of the "cell-soul," and both proceed from the "mechanical energy of the protoplasm." The sensibility of the irritable protoplasm is the same in the vegetable-cell of the Mimosa as in the animal-cell of the Hydra. How far Du Bois-Reymond is from discerning this, and how deeply he [107] is still entangled in neuro-psychological views is shown most clearly in the astonishing sentence which he has thought good to append to his above-quoted, erroneous assertion. "And what could we reply to the naturalist if, before he could agree to the assumption of a World-soul he required that we should show him—bedded in neuroglia and nourished by warm arterial blood—anywhere in the world a convolution of ganglionic centres co-extensive with the psychic capacity of such a Soul" (!)

In other respects we will not deny that Du Bois-Reymond stands far nearer to our recent evolution-theory than Virchow; nay, that from year to year he has always pronounced more and more emphatically in favour of the theory of descent as the one possible explanation of morphological phenomena; indeed, Du Bois-Reymond has lately counted himself as one of those naturalists who were convinced of the truth of evolution even before Darwin! Then it is only to be wondered why so acute and gifted an inquirer, who is certainly not lacking in scientific ambition, left it to Charles Darwin to place the egg of Columbus on the ring and to point out to biological science a new method of unlimited capacity by giving the theory of descent a definite and reliable basis!

It is clear from some remarks in his discourse bearing the title "Darwin versus Galiani" (1876), that Du [108] Bois-Reymond is still far from understanding the full significance of transmutation as affording a mechanical explanation of morphological problems. In this paper the "History of Creation" is treated simply as a romance, and the genealogies of phylogenesis are in his eyes "of about as much value as the pedigrees of the Homeric heroes are in the eyes of historical critics." Geologists may be extremely grateful for this estimate of their science, for undoubtedly geology, as a structure of hypotheses, is neither more nor less justifiable than phylogenesis, as I have already pointed out in my Munich address: "Our phylogenetic hypotheses may claim to have equal value with the universally-admitted hypotheses of geology; the only difference is this, that the mighty structure of hypotheses called geology is incomparably more complete, simpler, and easier to grasp than that more youthful one called phylogenesis." But as to the much-talked-of "genealogies," though they are nothing more than the simplest, barest, and most superficial expression of the hypotheses of phylogenesis, as provisional hypotheses they are just as indispensable to specific phylogenesis as the theoretical section-tables of the strata of the earth's crust are to geology.

If Du Bois-Reymond is so convinced of the truth of transmutation as he has lately given himself out to be, why does not he make at least one earnest [109] attempt to test the interpreting power of the theory of descent in physiology—his own most special province of inquiry? Why does he not labour at that hitherto quite unworked-out branch, physiogenesis, at the history of the evolution of functions, at the ontogenesis and phylogenesis of vital processes? The one idea which has lately been often spoken of as an important discovery of Du Bois-Reymond's—[the idea which had already been anticipated by Leibnitz, that the "innate ideas,"—intuitions *à priori*— have originated by transmission from primordial experience, *i.e.*, empirical, *à posteriori* convictions], was distinctly enunciated by me long before Du Bois-Reymond (as he omits to mention), in 1866, in my "General Morphology" (vol. ii. p. 446), and in 1868 in the "History of Creation" (vol. i. p. 31, vol. ii. p. 344). If Du Bois-Reymond had practically busied himself with these problems he would certainly have thought a little about the development of consciousness, and not have set down as an eternally insoluble problem, "How is it possible that matter can think?"—a form of words, be it observed, which has about as much sense as "how matter runs," or "how matter strikes the hours." Surely he would have guarded himself in that case from uttering the ponderous "Ignorabimus."

The question has been repeatedly asked why two such prominent Berlin biologists as Virchow and Du [110] Bois-Reymond availed themselves of the particularly solemn occasions of the fiftieth anniversary and of the fiftieth meeting of the German naturalists and physicians to lay lance in rest against the progress and freedom of science. The eager approbation which they both promptly met with from the party of the clergy and of all other enemies of free thought—Virchow, indeed, in much greater measure than Du Bois-Reymond—appears to justify this inquiry. I believe I can contribute something towards answering it, and as I am not fettered by any reverence for the Berlin tribunal of science or by any anxiety as to vexing influential Berlin connections, as most of my colleagues are who think as I do, I do not hesitate, here as elsewhere, to express my honest conviction in the freest and frankest manner, not troubling myself about the wrath which may be roused in many actual—and not actual—officials in Berlin at this exposition of the unvarnished truth.

The primary cause of their "misunderstanding," and the best excuse that can be offered for it, in Virchow and Du Bois-Reymond alike, lies in their unacquaintance with the advance of modern morphology. As has been repeatedly stated, no natural science is so directly to be referred to the doctrine of evolution—and more particularly to the theory of descent—as morphology. It is because we morphologists can [111] neither explain nor comprehend all the manifold and infinitely complex form-phenomena of the animal and plant worlds without this theory, because to us transmutation contains the only possible, rational explanation of organic types, that we all regard it as the indispensable basis of the scientific doctrine of form, and as demanding no further proofs of its certainty than those which now lie in abundance before us.

Du Bois-Reymond, and still more Virchow, ignore these proofs, because they are to a great extent ignorant alike of the inquiries and results, of the methods and the aims of our modern morphology, and this ignorance may be accounted for partly by the one-sided direction which their biological studies have taken, partly by the fact that there are few universities where the study of morphology is so behindhand as at the University of Berlin. Fully twenty years have now elapsed since the great Johannes Müller died, the last naturalist who could command all the departments of biology. The three great provinces of science which had been reunited into a triune kingdom under his powerful sceptre, were then divided among three professors' chairs: Du Bois-Reymond took that of physiology, Virchow, theoretical pathology (pathological anatomy and physiology), and the third, and most important chair, that of morphology (human and comparative anatomy, including the history of [112] evolution) fell to Boguslaus Reichert. This choice was, as is now universally admitted, an incomprehensible mistake. Instead of calling Carl Gegenbaur, or Max Schultze, or some one else of youthful capacity and vigour to the chair of morphology — a science which is the first foundation of zoology as well as of medicine — in Reichert they selected an elderly school anatomist cramped by strong old-fashioned notions, who had done some good and useful specialist work, but whose general views had developed all awry, and who for the unexampled obscurity of his conceptions and the confusion of his ideas, was outdone by none save only Adolf Bastian. For twenty years this man has represented animal morphology in the second university of Germany, and in these twenty years hardly any work worth mentioning has been done there in the whole of this vast department — neither by the master nor by his pupils. We have only to compare the many worthless anatomical productions of Berlin during these two decades (for instance, the recent confused work by Fritsch on the brain of fishes) with the rich mine of invaluable work produced during the preceding twenty years by Johannes Müller and his crowd of disciples.

But, as if this were not enough, Reichert took advantage of his influential position to hinder as far as possible all scientific study of morphology. For [113] example, he, with the co-operation of his colleagues, carried through that pretended "reform" of medical examination which puts the so-called *Tentamen physicum* in the place of the *philosophicum*; philosophy was entirely eliminated. Zoology and botany, which for centuries have been very justly regarded as the indispensable foundation of all instruction in natural science for the young medical student, disappeared from the curriculum. Only, as if in scorn of these sciences, in each examination a small place was reserved for comparative anatomy—for that most difficult and philosophical part of animal morphology which cannot be at all understood without some previous knowledge of the other branches of zoology. And yet comparative anatomy and the history of development are the indispensable preliminary steps to a true scientific comprehension of human anatomy, that most essential foundation of all medical knowledge. Without the vivifying idea of development, mere anatomical knowledge is an empty and lifeless cramming of the memory.

In the place of morphology, thus degraded from its office, a detailed study of physiology was introduced, but always in a one-sided direction. Now these two great branches of biology, which are equally important and have an equal claim on our attention, are so dependent the one on the other, that a real scientific [114] understanding of organic life can never be obtained without due relative study of both. The masterly and incomparable teaching of Johannes Müller owed a great part of its captivating charm to his equitable regard for morphology and physiology, as well as to his comprehensive treatment, from the broadest point of view, of the enormous mass of details to be dealt with. I therefore have not the smallest doubt that the morphological training of medical students, as at present conducted at Berlin under the influence of Reichert and his colleagues, is as far behind that of Müller's day, twenty or thirty years ago, in all general comprehension of the typical organism, as it is in advance of it in specialist acquirements.

In medical, as in all other scientific learning, the highest aim does not consist in seeking to accumulate a vast chaotic mass of isolated items of knowledge, but in a general comprehension of the science, its aims and problems. The teacher should, above everything, guide the pupil to this general knowledge, and then it will be easy to him, by the aid of proper methods, to acquire mastery in each individual and special branch. Thus in medicine, as in every other science, he is not the best qualified who, on Bastian's method, has loaded his memory with a confused mass of undigested facts, and has flung them all together into his brain without any order; but, on the contrary, he who has practically [115] digested a considerable number of the most important facts, and has critically co-ordinated them to a harmonious whole. It is precisely under this aspect that transmutation is of such inestimable value to morphology; it enables us to rise from the bare empirical knowledge of numberless isolated facts to a philosophical conception of their efficient causes.

The aversion and contempt which the theories of descent and selection have met with at Berlin, more than in any other place, is in great measure to be explained by the circumstance that, during the last two decades, morphological studies have been more neglected in that university than any others. In no other city of Germany has evolution in general, as well as Darwinism in particular, been so little valued, so utterly misunderstood, and treated with such sovereign disdain as in Berlin. Nay, Adolf Bastian, the most zealous of all the Berlin opponents of our doctrines, has insisted on these facts with peculiar satisfaction. Of all the conspicuous naturalists of Berlin only one accepted the doctrine of transmutation from the beginning with sincere warmth and full conviction, being, indeed, persuaded of its truth even before Darwin himself. This was the gifted botanist Alexander Braun, who is lately dead—a morphologist who was equally distinguished by the extent of his comprehensive knowledge of details, as by his philosophical [116] mastery over them. His firm conviction of the truth of the theory of descent is all the more remarkable because he was at the same time a spotless character, a pious Christian in the best sense of the word, and an extremely conservative politician; a striking example that these convictions can dwell side by side with the principles of the recent doctrines of evolution in one and the same person. But in comparison with the powerful influence of the rest of the Berlin naturalists who, for the most part, are decided opponents of transmutation, and who have only lately—a few of them, to follow the fashion—become converts to it, a man like Alexander Braun could have no effect in procuring that it should be taught.

However, this is not the first time that this very Berlin society of learned men has set itself with remarkable firmness against the most important advances of science. Virchow's former colleague, the deceased Stahl, with a similar purpose and with great success, preached this principle: "Science must turn back again." Just as at the present day the Berlin biologists have opposed the most obstinate and pertinacious resistance to the greatest scientific stride of this century, so did it happen in former times with regard to other doctrines of progress. We have only to recall Caspar Friedrich Wolff, the great inquirer, [117] who in 1759 first detected the nature of the individual processes of development in the animal ovum, and founded on it his observations in his "Theoria Generationes," which marked an epoch in biological science. The Berlin savants, full of the prevailing prejudices, so contrived at that time that Wolff never once could obtain the permission which he craved, to lecture publicly, and in consequence found himself compelled to retire to St. Petersburg for the sake of peace. And yet in that instance there was no question of a "theory" properly so-called. For the fundamental theory of generation—the "theory of epigenesis"—as propounded by Wolff was nothing more than a simple, general exposition of embryological facts which he had been the first to recognise, and of whose truth every one might convince himself by direct observation. In spite of this, for another half century, the predominant error of the "Preformation-theory" continued to be universally accepted— the ludicrous and nonsensical doctrine, supported by the authority of Haller, that all the successive generations of animals exist preconceived and enclosed one within the other, and that no individual development ever takes place! *Nulla est epigenesis!* (Compare my "Evolution of Man," vol. i. p. 31.)

But it would appear that it is the fate of that most [118] interesting of all sciences, the history of evolution, to find its most important steps and its greatest discoveries met by the firmest and most persistent opposition. For while Wolff's fundamental theory of epigenesis, which was promulgated in 1759, was not recognised until 1812, Lamarck's theory of descent, founded in 1809, had to wait fully fifty years before Darwin, in 1859, showed it to be the greatest acquisition of modern science; and during that period, in spite of all the progress made in empirical science, how persistently this most comprehensive of all biological theories was combated. We need only recollect how, in 1830, the celebrated George Cuvier silenced its most eloquent supporter, Geoffroy St. Hilaire, in the midst of the Paris Academy, and how almost at the same time its founder, the great Lamarck, ended his life in blindness, misery and want, while his opponent Cuvier was enjoying the highest honours and the greatest splendour. And yet we know now that the despised and contemned Lamarck and Geoffroy had already grasped truths of the highest significance, while Cuvier's much-admired and universally-accepted theory of creation is now on all hands neglected as an absurd and untenable delusion. But as neither Haller as against Wolff, nor Cuvier as against Lamarck, could permanently hinder the [119] progress of free inquiry, neither will Virchow succeed in turning back the course of Darwin's admirable achievement; no, not even when he is supported by the discourses of his friend Bastian.

While we cannot but earnestly lament Virchow's inimical attitude in this great struggle for truth, we must not overlook the effects of his well-founded authority in a yet wider sphere. For instance, the hostile attitude which the greater part of the Berlin press persistently maintains towards the doctrine of development (particularly the Liberal "National-Zeitung") is to be referred to the influence of his authority. But much as this reactionary vein, in this and in other intelligent circles at Berlin, must be regretted on the one hand, on the other we must observe that by this evil we have been preserved from a far greater one. This greater evil — the greatest, in fact, which German science could have to encounter — would be the monopoly of knowledge at Berlin; a Centralisation of Science. The injurious fruits of this system of centralisation in France, for instance, the continual deterioration of French science through the Parisian "Monopoly of Knowledge," and its steady decline during half a century from the sublimest heights — these are all well known. From such a centralisation of German science — which would be especially dangerous if it occurred in the capital, [120] Berlin — we may hope to be preserved; in the first place by the manifold differences and the many-sided individuality of the German national spirit, the much-abused German provincialism (Particularismus). While these provincial modes of thought can never have any permanent political value, nor be productive of a desirable form of government, it is beyond a doubt that their outcome has been fruitful and happy for German science. For it owes its splendid pre-eminence over that of other countries precisely to the many centres of culture which were offered by those numerous petty capitals of the minor German

States which strove to outdo each other in eager emulation. It is to be hoped that this happy decentralisation of science in our politically united fatherland may continue to subsist!

And next to this centrifugal tendency of our German national mind nothing will so greatly contribute to it as a vigorous opposition to the free advance of science, such as is just now declaring itself in the metropolis. For by just so much as Berlin is dragged back by it in the mighty onward stream of free intellectual movement, by so much will it see itself outstripped by the other seats of culture in Germany, which follow the stream with enthusiasm, or at least without resistance. If Emil du Bois-Reymond raises [121] the cry of "Ignorabimus," and Rudolf Virchow his still more audacious one of "Restringamur," as the watchwords of science, then, from Jena, let the shout be raised and echoed from a hundred other universities — "Impavidi progrediamur!"

THE END.

WORKS OF PROFESSOR ERNST HAECKEL.

FREEDOM IN SCIENCE AND TEACHING. From the German of Ernst Haeckel. With a Prefatory Note by T. H. Huxley, F.R.S. 1 vol., 12mo.

THE EVOLUTION OF MAN. A Popular Exposition of the Principal Points of Human Ontogeny and Phylogeny. From the German of Ernst Haeckel, Professor in the University of Jena, author of "The History of Creation," etc. With numerous Illustrations. In two vols., 12mo. Cloth. Price, $5.00.

From the London Saturday Review.

"In this excellent translation of Professor Haeckel's work, the English reader has access to the latest doctrines of the Continental school of evolution, in its application to the history of man. It is in Germany, beyond any other European country, that the impulse given by Darwin twenty years ago to the theory of evolution has influenced the whole tenor of philosophical opinion. There may be, and are, differences in the degree to which the doctrine may be held capable of extension into the domain of mind and morals; but there is no denying, in scientific circles at least, that as regards the physical history of organic nature much has been done toward making good a continuous scheme of being."

THE HISTORY OF CREATION; or, the Development of the Earth and its Inhabitants by the Action of Natural Causes. A Popular Exposition of the Doctrine of Evolution in general, and of that of Darwin, Goethe, and Lamarck in particular. From the German of Ernst Haeckel, Professor in the University of Jena. The translation revised by Professor E. Ray Lankester, M.A., F.R.S., Fellow of Exeter College, Oxford. Illustrated with Lithographic Plates. In 2 vols., 12mo. Cloth, $5.00.

WORKS OF THOMAS H. HUXLEY, LL. D., F.R.S.

MAN'S PLACE IN NATURE. 1 vol., 12mo. Cloth, $1.25.

LAY SERMONS, ADDRESSES, AND REVIEWS. 1 vol., 12mo. Cloth, $1.75.

ON THE ORIGIN OF SPECIES. 1 vol., 12mo. Cloth, $1.00.

CRITIQUES AND ADDRESSES. 12mo. Cloth, $1.50.

MORE CRITICISMS ON DARWIN, AND ADMINISTRATIVE NIHILISM. 1 vol., 12mo. Limp cloth, 50 cents.

AMERICAN ADDRESSES; with a Lecture on the Study of Biology. 12mo. Cloth, $1.25.

A MANUAL OF THE ANATOMY OF VERTEBRATED ANIMALS. Illustrated. 1 vol., 12mo. Cloth, $2.50.

PHYSIOGRAPHY: an Introduction to the Study of Nature. With Illustrations and Colored Plates. 12mo. Cloth, $2.50.

A MANUAL OF THE ANATOMY OF INVERTEBRATED ANIMALS. Illustrations and Colored Plates. 12mo. Cloth, $2.50.

ELEMENTS OF PHYSIOLOGY AND HYGIENE. By T. H. Huxley and W. J. Youmans. 1 vol., 12mo. $1.50.

RECENT EDUCATIONAL WORKS.

Principles and Practice of Teaching.

By JAMES JOHONNOT.

1 volume, 12mo. Cloth. 396 pages. Price, $1.50.

CONTENTS.

I. What is Education?

II. The Mental Powers: their Order of Development, and the Methods most conducive to Normal Growth.

III. Objective Teaching: its Methods, Aims, and Principles.

IV. Subjective Teaching: its Aims and Place in the Course of Instruction.

V. Object-Lessons: their Value and Limitations.

VI. Relative Value of the Different Studies in a Course of Instruction.

VII. Pestalozzi, and his Contributions to Educational Science.

VIII. Froebel and the Kindergarten.

IX. Agassiz; and Science in its Relation to Teaching.

X. Contrasted Systems of Education.

XI. Physical Culture.

XII. Æsthetic Culture.

XIII. Moral Culture.

XIV. A Course of Study.

XV. Country Schools.

Extract from Preface.

"Experience is beginning to show that teaching, like every other department of human thought and activity, must change with the changing conditions of society, or it will fall in the rear of civilization, and become an obstacle to improvement…. In this volume an endeavor has been made to examine education from the standpoint of modern thought, and to contribute something to the solution of the problems that are forcing themselves upon the attention of educators. To these ends, a concise statement of the well-settled principles of psychology has been made, and a connected view of the interdependence of the sciences given, to serve as a guide to methods of instruction, and to determine the subject-matter best adapted to each stage of development. The systems of several of the great educational reformers have been analyzed, with a view to ascertain precisely what each has contributed to the science of teaching, and how far their ideas conform to psychological laws; and an endeavor has been made to combine the principles derived from both experience and philosophy into one coherent system."

ELEMENTARY LESSONS IN HISTORICAL ENGLISH GRAMMAR, containing Accidence and Word-Formation. By the Rev. Richard Morris, LL. D., President of the Philological Society, London. 18mo. Cloth, 254 pages. Price, $1.00.

WORDS, and how to put them together. By Harlan H. Ballard, Principal of Lenox High-School, Lenox, Mass. 18mo. Cloth. Price, 40 cents.

GENERAL HISTORY, from b. c. 800 to a. d. 1876. Outlined in Diagrams and Tables; with Index and Genealogies. For General Reference, and for Schools and Colleges. By Samuel Willard, A. M., M. D., Professor of History in Chicago High-School. 8vo. Cloth. Price, $2.00.

HARKNESS'S PREPARATORY COURSE IN LATIN PROSE AUTHORS, comprising four books of Cæsar's Gallic War, Sallust's Catiline, and eight Orations of Cicero. With Notes, Illustrations, a Map of Gaul, and a Special Dictionary. 12mo. Cloth. Price, $1.75.

HARKNESS'S SALLUST'S CATILINE, with Notes and a Special Vocabulary. 12mo. Cloth. Price, $1.15.

THE LATIN SPEAKER. Easy Dialogues, and other Selections for Memorizing and Declaiming in the Latin Language. By Frank Sewall, A. M. 12mo. Cloth. Price, $1.00.

New Volume of "The International Scientific Series."

EDUCATION AS A SCIENCE.

BY

ALEXANDER BAIN, LL. D.,

PROFESSOR OF LOGIC IN THE UNIVERSITY OF ABERDEEN.

1 *vol.*, 12*mo. Cloth, price*, $1.75.

"In the present work I have surveyed the Teaching Art, as far as possible, from a scientific point of view; which means, among other things, that the maxims of ordinary experience are tested and amended by bringing them under the best ascertained laws of the mind." — *From Preface.*

"Dr. Bain's *renovated curriculum* is certainly extensive enough, even if it omits Greek and Latin. According to this, higher education should embrace—first, science; second, the humanities, including history and the social science, and some portions of the universal literature; and, third, English composition and literature."—*New York Evening Express.*

"The work should become a text-book for teachers, not to be followed servilely or thoughtlessly, but used for its suggestiveness."—*Boston Gazette.*

"Professor Bain is not a novice in this field. His work is admirable in many respects for teacher, parent, and pupil."—*Philadelphia North American.*

"A work of great value to all teachers who study it intelligently."—*Boston Advertiser.*

"At once speculative and practical, entering largely into the philosophy of teaching, and manfully handling facts."—*Philadelphia Press.*

PRIMERS

IN SCIENCE, HISTORY and LITERATURE.

18mo. Flexible cloth, 45 cents each.

I. — Edited by Professors Huxley, Roscoe, and Balfour Stewart.

SCIENCE PRIMERS.

Chemistry ... H. E. Roscoe.

Botany ... J. D. Hooker.

Physics ... Balfour Stewart.

Logic ... W. S. Jevons.

Physical Geograph ...y A. Geikie.

Inventional Geology ... A. Geikie.

Geometry ... W. G. Spencer.

Physiology ... M. Foster.

Pianoforte ... Franklin Taylor.

Astronomy ... J. N. Lockyer.

Political Economy ... W. S. Jevons.

II. — Edited by J. R. Green, M.A., *Examiner in the School of Modern History at Oxford.*

HISTORY PRIMERS.

Greece ... C. A. Fyffe.

Old Greek Life ... J. P. Mahaffy.

Rome ... M. Creighton.

Roman Antiquities ... A. S. Wilkins.

Europe ... E. A. Freeman.

Geography ... George Grove.

III. — Edited by J. R. Green, M.A.

LITERATURE PRIMERS.

English Grammar ... R. Morris.

Shakespeare ... E. Dowden.

English Literature ... Stopford Brooke.

Studies in Bryant ... J. Alden.

Philology ... J. Peile.

Greek Literature ... R. C. Jebb.

Classical Geography ... M. F. Tozer

English Grammar Exercises ... R. Morris.

Homer ... W. E. Gladstone.

(Others in preparation.)

The object of these primers is to convey information in such a manner as to make it both intelligible and interesting to very young pupils, and so to discipline their minds as to incline them to more systematic after-studies. They are not only an aid to the pupil, but to the teacher, lightening the task of each by an agreeable, easy, and natural method of instruction. In the Science Series some simple experiments have been devised, leading up to the chief truths of each science. By this means the pupil's interest is excited, and the memory is impressed so as to retain without difficulty the facts brought under observation. The woodcuts which illustrate these primers serve the same purpose, embellishing and explaining the text at the same time.

Appletons' School Readers,

CONSISTING OF FIVE BOOKS.

BY

| W. T. HARRIS, LL. D., *Superintendent of Schools, St. Louis, Mo.* | A. J. RICKOFF, A. M., *Superintendent of Instruction, Cleveland, O.* | MARK BAILEY, A. M., *Instructor in Elocution, Yale College.* |

Retail Prices.

APPLETONS' FIRST READER	$0.25
APPLETONS' SECOND READER	.40
APPLETONS' THIRD READER	.52
APPLETONS' FOURTH READER	.70
APPLETONS' FIFTH READER	1.25

CHIEF MERITS.

These Readers, while avoiding extremes and one-sided tendencies, combine into one harmonious whole the several results desirable to be attained in a series of school reading-books. These include good pictorial illustrations, a combination of the word and phonic methods, careful grading, drill on the peculiar combinations of letters that represent vowel-sounds, correct spelling, exercises well arranged for the pupil's preparation by himself (so that he shall learn the great lessons of self-help, self-dependence, the habit of application), exercises that develop a practical command of correct forms of expression, good literary taste, close critical power of thought, and ability to interpret the entire meaning of the language of others.

THE AUTHORS.

The high rank which the authors have attained in the educational field and their long and successful experience in practical school-work especially fit them for the preparation of text-books that will embody all the best elements of modern educative ideas. In the schools of St. Louis and Cleveland, over which two of them have long presided, the subject of reading has received more than usual attention, and with results that have established for them a wide reputation for superior elocutionary discipline and accomplishments. Feeling the need of a series of reading-books harmonizing in all respects with the modes of instruction growing out of their long tentative work, they have carefully prepared these volumes in the belief that the special features enumerated will commend them to practical teachers everywhere.

Of Professor Bailey, Instructor of Elocution in Yale College, it is needless to speak, for he is known throughout the Union as being without a peer in his profession. *His methods make natural, not mechanical readers.*

A SHORT HISTORY

OF

Natural Science and the Progress of Discovery,

FROM THE TIME OF THE GREEKS TO THE PRESENT DAY.

FOR SCHOOLS AND YOUNG PERSONS.

By ARABELLA B. BUCKLEY.
With Illustrations. 12mo.   Cloth, $2.00.

"During many years the author acted as secretary to Sir Charles Lyell, and was brought in contact with many of the leading scientific men of the day, and felt very forcibly how many important facts and generalizations of science, which are of great value both in the formation of character and in giving a true estimate of life and its conditions, are totally unknown to the majority of otherwise well-educated persons. This work has been written for this purpose, and it is not too much to say that it will effect its purpose."—*European Mail.*

"The volume is attractive as a book of anecdotes of men of science and their discoveries. Its remarkable features are the sound judgment with which the true landmarks of scientific history are selected, the conciseness of the information conveyed, and the interest with which the whole subject is nevertheless invested. Its style is strictly adapted to its avowed purpose of furnishing a text-book for the use of schools and young persons."—*London Daily News.*

"Before we had read half-a-dozen pages of this book we laid it down with an expression of admiration of the wonderful powers of the writer. And our opinion has increased in intensity as we have gone on, till we have come to the conclusion that it is a book worthy of being ranked with Whewell's 'History of the Inductive Sciences'; it is one which should be first placed in the hands of every one who proposes to become a student of natural science, and it would be well if it were adopted as a standard volume in all our schools."—*Popular Science Review.*

"A most admirable little volume. It is a classified résumé of the chief discoveries in physical science. To the young student it is a book to open up new worlds with every chapter."—*Graphic.*

"We have nothing but praise for this interesting book. Miss Buckley has the rare faculty of being able to write for young people." — *London Spectator.*

"The book will be a valuable aid in the study of the elements of natural science." — *Journal of Education.*

FAIRY-LAND OF SCIENCE.

BY ARABELLA B. BUCKLEY,

Author of "A Short History of Natural Science," etc.

WITH NUMEROUS ILLUSTRATIONS.

12mo   Cloth, price, $1.50.

"A child's reading-book admirably adapted to the purpose intended. The young reader is referred to nature itself rather than to books, and is taught to observe and investigate, and not to rest satisfied with a collection of dull definitions learned by rote and worthless to the possessor. The present work will be found a valuable and interesting addition to the somewhat overcrowded child's library." — *Boston Gazette.*

"Written in a style so simple and lucid as to be within the comprehension of an intelligent child, and yet it will be found entertaining to maturer minds." — *Baltimore Gazette.*

"It deserves to take a permanent place in the literature of youth." — *London Times.*

"The ease of her style, the charm of her illustrations, and the clearness with which she explains what is abstruse, are no doubt the result of much labor; but there is nothing labored in her pages, and the reader must be dull indeed who takes up this volume without finding much to attract attention and to stimulate inquiry." — *Pall Mall Gazette.*

"So interesting that having once opened it we do not know how to leave off reading." — *Saturday Review.*

"We are compelled to admit that there is indeed a fairy-land of science. This is the fairy-land upon which Miss Arabella Buckley lectured last year, and upon which she has now produced a child's reading-book, which is most charmingly illustrated, and which is in every way rendered especially interesting to the juvenile reader." — *London Athenæum.*

THE

Experimental Science Series.

In neat 12mo volumes, bound in cloth, fully illustrated. Price per volume, $1.00.

This series of scientific books for boys, girls, and students of every age, was designed by Prof. Alfred M. Mayer, Ph. D., at the Stevens Institute of Technology, Hoboken, New Jersey. Every book is addressed directly to the young student, and he is taught to construct his own apparatus out of the cheapest and most common materials to be found. Should the reader make all the apparatus described in the first book of this series, he will spend only $12.40.

NOW READY:

I.—LIGHT.

A Series of Simple, Entertaining, and Inexpensive Experiments in the Phenomena of Light, for Students of every Age.

By *ALFRED M. MAYER and CHARLES BARNARD.*

II.—SOUND.

A Series of Simple, Entertaining, and Inexpensive Experiments in the Phenomena of Sound, for the Use of Students of every Age.

By *ALFRED MARSHALL MAYER,*

Professor of Physics in the Stevens Institute of Technology; Member of the National Academy of Sciences; of the American Philosophical Society, Philadelphia; of the American Academy of Arts and Sciences, Boston; of the New York Academy of Sciences; of the German

Astronomical Society; of the American Otological Society; and Honorary Member of the New York Ophthalmological Society.

In Active Preparation:

III. Vision and the Nature of Light.
IV. Electricity and Magnetism.
V. Heat.
VI. Mechanics.
VII. Chemistry.
VIII. The Art of experimenting with Cheap and Simple Instruments.

LIGHT:

A Series of Simple, Entertaining, and Inexpensive Experiments in the Phenomena of Light, for the Use of Students of Every Age.

BY ALFRED M. MAYER and CHARLES BARNARD.

Neat 12mo volume, fully illustrated.   Cloth, price, $1.00.

"Professor Mayer has invented a series of experiments in Light which are described by Mr. Barnard. Nothing is more necessary for sound-teaching than experiments made by the pupil, and this book, by considering the difficulty of costly apparatus, has rendered an important service to teacher and student alike. It deals with the sources of light, reflection, refraction, and decomposition of light. The experiments are extremely simple and well suited to young people." — *Westminster Review.*

"This work describes, in simple language, a number of experiments illustrating the principal properties of light, by means of a beam of sunlight admitted into a dark room, and various contrivances. The experiments are highly ingenious, and the young student can not fail to learn a great deal from the book. As an example of the effective experimental method employed, we may specially mention the device for illustrating the refraction of light. This book is specially designed 'to give to every teacher and scholar the knowledge of the art of experimenting.'" — *The Quarterly Journal of Science* (London).

"A singularly excellent little hand-book for the use of teachers, parents, and children. The book is admirable both in design and execution. The experiments for which it provides are so simple that an intelligent boy or girl can easily make them, and so beautiful and interesting that even the youngest children must enjoy the exhibition. The experiments here described are abundantly worth all that they cost in money and time in any family where there are boys and girls to be entertained." — *New York Evening Post.*

"The experiments are capitally selected, and equally as well described. The book is conspicuously free from the multiplicity of confusing directions with which works of the kind too often abound. There is an abundance of excellent illustrations." — *New York Scientific American.*

"The experiments are for the most part new, and have the merit of combining precision in the methods with extreme simplicity and elegance of design. The value of the book is further enhanced by the numerous carefully-drawn cuts, which add greatly to its beauty." — *American Journal of Science and Arts.*

SOUND:

A Series of Simple, Entertaining, and Inexpensive Experiments in the Phenomena of Sound, for the Use of Students of Every Age.

By ALFRED MARSHALL MAYER,

Professor of Physics in the Stevens Institute of Technology; Member of the National Academy of Sciences, etc.

Uniform with "LIGHT," first volume of the Series.

Neat 12mo volume, fully illustrated.   Cloth, price, $1.00.

"It would be difficult to find a better example of a series which is excellent throughout. This little work is accurate in detail, popular in style, and lucid in arrangement. Every statement is accompanied with ample illustrations. We can heartily recommend it, either as an introduction to the subject or as a satisfactory manual for those who have no time for perusing a larger work. It contains an excellent description, with diagrams, of Faber's Talking Machine and of Edison's Talking Phonograph, which can not fail to be interesting to any reader who takes an interest in the marvelous progress of natural science." — *British Quarterly.*

"The style of the book is very clear, and the experiments interesting. It can not fail to have an important educational influence." — *Westminster Review.*

"It would really be difficult to exaggerate the merit, in the sense of consummate adaptation to its modest end, of this little treatise on 'Sound.' It teaches the youthful student how to make experiments for himself, without the help of a trained operator, and at very little expense. These hand-books of Professor Mayer should be in the hands of every teacher of the young." — *New York Sun.*

"An admirably clear and interesting collection of experiments, described with just the right amount of abstract information and no more, and placed in progressive order. The recent inventions of the phonograph and microphone lend an extraordinary interest to this whole field of experiment, which makes Professor Mayer's manual especially opportune." — *Boston Courier.*

The Works of Professor E. L. YOUMANS, M. D.

Class-book of Chemistry.

New edition. 12mo. Cloth, $1.50.

The Hand-book of Household Science.

A Popular Account of Heat, Light, Air, Aliment, and Cleansing, in their Scientific Principles and Domestic Applications. 12mo. Illustrated. Cloth, $1.75.

The Culture demanded by Modern Life.

A Series of Addresses and Arguments on the Claims of Scientific Education. Edited, with an Introduction on Mental Discipline in Education. 1 vol., 12mo. Cloth, $2.00.

Correlation and Conservation of Forces.

A Series of Expositions by Professor Grove, Professor Helmholtz, Dr. Mayer, Dr. Faraday, Professor Liebig, and Dr. Carpenter. Edited, with an Introduction and Brief Biographical Notices of the Chief Promoters of the New Views, by Edward L. Youmans, M. D. 1 vol., 12mo. Cloth, $2.00.

The Popular Science Monthly.

Conducted by E. L. and W. J. Youmans.

Containing instructive and interesting articles and abstracts of articles, original, selected, and illustrated, from the pens of the leading scientific men of different countries;

Accounts of important scientific discoveries;

The application of science to the practical arts;

The latest views put forth concerning natural phenomena, by savants of the highest authority.

TERMS: Five dollars per annum; or fifty cents per number. A Club of five will be sent to any address for $20.00 per annum.

The volumes begin May and November of each year. Subscriptions may begin at any time.

THE POPULAR SCIENCE MONTHLY and APPLETONS' JOURNAL, together, for $7.00 per annum, postage prepaid.

A New and Valuable Work for the Practical Mechanic and Engineer.

APPLETONS'

CYCLOPÆDIA OF APPLIED MECHANICS.

A DICTIONARY OF

MECHANICAL ENGINEERING AND THE MECHANICAL ARTS.

ILLUSTRATED BY 5,000 ENGRAVINGS.

Edited by **PARK BENJAMIN, Ph. D.**

CONTRIBUTORS.

T. A. EDISON, Ph. D.	ABRAM L. HOLLEY, C. E.
RICHARD H. BULL, C. E.	COLEMAN SELLERS, M. E.
SAMUEL WEBBER, C. E.	Prof. C. W. McCord.
CHARLES E. EMERY, C. E.	IRVING M. SCOTT, Esq.
Prof. DE VOLSON WOOD.	F. A. McDowell, C. E.
JOSHUA ROSE, M. E.	H. A. MOTT, Jr., Ph. D.
PIERRE de P. RICKETTS, Ph. D.	W. H. PAYNE, C. E.
Hon. ORESTES CLEVELAND.	GEORGE H. BENJAMIN, M. D.
W. T. J. KRAJEWSKI, C. E.	THERON SKELL, C. E.
S. W. GREEN, Esq.	WILLIAM KENT, C. E.
JOHN BIRKINBINE, C. E.	W. E. KELLY, Esq.
HENRY L. BREVOORT, C. E.	F. T. THURSTON, C. E.
Lieut. A. A. BOYD, U. S. N.	JOHN HOLLINGSWORTH, Esq.

Appletons' Cyclopædia of Applied Mechanics of 1879 is a new work, and not a revision of the former Dictionary of Mechanics of 1850. It aims to present the best and latest American practice in the mechanical arts, and to compare the same with that of other nations. It also exhibits the extent to which American invention and discovery have contributed to the world's progress during the last quarter century. Its production is deemed timely in view of the existing popular interest in the labors of the mechanic and inventor which has been awakened by the great International Expositions of the last decade, and by the wonderful discoveries made by American inventors during the past three years.

The Contributors whose names are given above number many of the most eminent American mechanical experts and engineers. Several of their contributions contain the results of original research and thought, never before published. Their efforts have in all cases tended to simplify the subjects treated, to avoid technicalities, and so to render all that is presented easily understood by the general reader as well as by the mechanical student. Examples are appended to all rules, explanations to all tables, and in such matters as the uses of tools and management of machines the instructions are unusually minute and accurate.

In semi-monthly Parts, 50 cents each.
Subscriptions received only for the entire work of Twenty-four Parts.

D. APPLETON & CO., Publishers, 549 & 551 Broadway, New York.

www.ingramcontent.com/pod-product-compliance
Lightning Source LLC
Chambersburg PA
CBHW031439210526
45464CB00005B/2265